Lecture Notes in Physics

T0238979

Volume 891

The Lecture Notes in Physics

The series Lecture Notes in Physics (LNP), founded in 1969, reports new developments in physics research and teaching-quickly and informally, but with a high quality and the explicit aim to summarize and communicate current knowledge in an accessible way. Books published in this series are conceived as bridging material between advanced graduate textbooks and the forefront of research and to serve three purposes:

- to be a compact and modern up-to-date source of reference on a well-defined topic
- to serve as an accessible introduction to the field to postgraduate students and nonspecialist researchers from related areas
- to be a source of advanced teaching material for specialized seminars, courses and schools

Both monographs and multi-author volumes will be considered for publication. Edited volumes should, however, consist of a very limited number of contributions only. Proceedings will not be considered for LNP.

Volumes published in LNP are disseminated both in print and in electronic formats, the electronic archive being available at springerlink.com. The series content is indexed, abstracted and referenced by many abstracting and information services, bibliographic networks, subscription agencies, library networks, and consortia.

Proposals should be sent to a member of the Editorial Board, or directly to the managing editor at Springer:

Christian Caron
Springer Heidelberg
Physics Editorial Department I
Tiergartenstrasse 17
69121 Heidelberg/Germany
christian.caron@springer.com

More information about this series at
http://www.springer.com/series/5304

Francesco Iachello

Lie Algebras
and Applications

Second Edition

 Springer

Francesco Iachello
Department of Physics
Yale University
P.O. Box 208120
New Haven, CT 06520-8120, USA

ISSN 0075-8450   ISSN 1616-6361 (electronic)
Lecture Notes in Physics
ISBN 978-3-662-44493-1   ISBN 978-3-662-44494-8 (eBook)
DOI 10.1007/978-3-662-44494-8
Springer Heidelberg New York Dordrecht London

Library of Congress Control Number: 2014951341

Printed on acid-free paper

Springer is part of Springer Science+Business Media (www.springer.com)

Preface

In the second part of the twentieth century, algebraic methods have emerged as a powerful tool to study theories of physical phenomena, especially those of quantal systems. The framework of Lie algebras, initially introduced by Sophus Lie in the last part of the nineteenth century, has been considerably expanded to include graded Lie algebras, infinite-dimensional Lie algebras, and other algebraic constructions. Algebras that were originally introduced to describe certain properties of a physical system, in particular behavior under rotations and translations, have now taken the center stage in the construction of physical theories.

This book contains a set of notes from lectures given at the Yale University and other universities and laboratories in the last 20 years. The notes are intended to provide an introduction to Lie algebras at the level of a one-semester graduate course in physics. Lie algebras have been particularly useful in spectroscopy, where they were introduced by Eugene Wigner and Giulio Racah. Racah's lectures were given at Princeton University in 1951 (Group Theory and Spectroscopy) and they provided the impetus for the initial applications in atomic and nuclear physics. In the intervening years, many other applications have been made. This book contains a brief account of some of these applications to the fields of molecular, atomic, nuclear, and particle physics. The application of Lie algebraic methods in physics is so wide that often students are overwhelmed by the sheer amount of material to absorb. This book is intended to give a basic introduction to the method and how it is applied in practice, with emphasis to bosonic systems as encountered in molecules (vibron model), and in nuclei (interacting boson model), and to fermionic systems as encountered in atoms (atomic shell model), in nuclei (nuclear shell model), and in hadrons (quark model). Exactly solvable problems in quantum mechanics are also discussed.

Associated with a Lie algebra there is a Lie group. Although the emphasis of these lecture notes is on Lie algebras, a chapter is devoted to Lie groups and to the relation between Lie algebras and Lie groups.

Many exhaustive books exist on the subject of Lie algebras and Lie groups. Reference to these books is made throughout, so that the interested student can study the subject in depth. A selected number of other books, not explicitly mentioned in the text, are also included in the reference list, to serve as additional introductory material and for cross-reference.

In the early stages of preparing the notes, I benefited from many conversations with Morton Hamermesh, Brian Wybourne, Asim Barut, and Jin-Quan Chen, who wrote books on the subject, but are no longer with us. This book is dedicated to their memory. I also benefited from conversations with Robert Gilmore, who has written a book on the subject, and with Phil Elliott, Igal Talmi, Akito Arima, Bruno Gruber, Arno Böhm, Yuval Ne'eman, Marcos Moshinsky, Yuri Smirnov, and David Rowe, who have made major contributions to this field.

I am very much indebted to Mark Caprio for a critical reading of the manuscript, and for his invaluable help in preparing the final version of these lecture notes.

New Haven, CT, USA   Francesco Iachello
May 2006

Preface to the Second Edition

Algebraic methods continue to be a powerful tool to study theories of physical phenomena. This second edition contains an enlarged set of notes from lectures given at the Yale University and other universities and laboratories in the last 6 years. Relatively to the first edition, three new chapters have been added, discussing contractions of Lie algebras and introducing the concept of globally symmetric Riemannian spaces (coset spaces). Explicit construction of some of these spaces, especially those of interest in the study of bosonic systems and in quantum mechanics, is included.

The notes are still intended to provide an introduction to Lie algebras and applications at the level of a one-semester graduate course in physics. For this reason, the additional material has been kept brief and at the same level of the original material.

References to mathematics books where the interested reader can find details on the somewhat elaborate constructions and proofs of Lie theory have also been included.

Phil Elliott, Yuval Ne'eman, Marcos Moshinsky, and Larry Biedenharn, with whom I had many conversations in preparing the first edition of this book, are no longer with us. This second edition is dedicated to them.

I am very much indebted to Jenni-Mari Kotila for her help in preparing the final version of these expanded lecture notes.

New Haven, CT, USA   Francesco Iachello
January 2014

Contents

List of Symbols

$\mathcal{A}$	Abelian algebra
$[,]_+$	Anticommutator
$\{,\}$	Anticommutator
$\mathcal{B}$	Basis (bosons)
$\mathcal{F}$	Basis (fermions)
$\in$	Belongs to
b	Boson annihilation operator
$b^\dagger$	Boson creation operator
$\langle|$	Bra
$C_i(g)$	Casimir operator of g
$[,]$	Commutator
C	Complex
$\supset$	Contains
$\subset$	Contained in
$\longrightarrow_c$	Contraction
G/H	Coset space
$\equiv$	Defined as
Der	Derivation
dim	Dimension of the representation
$\oplus$	Direct sum
DS	Dynamic symmetry
X_α	Element of the algebra
$\square$	Entry in the Young tableau
$\doteq$	Equal to
$\langle\rangle$	Expectation value

$\downarrow$ exp	Exponential map
a	Fermion annihilation operator
$a^{\dagger}$	Fermion creation operator
$\backslash/$	Gel'fand pattern
$\approx$	Homomorphic groups
$\cap$	Intersect
IRB	Irreducible basis
$\sim$	Isomorphic algebras
$\langle|\rangle$	Isoscalar factors
$|\rangle$	Ket
$\downarrow$	Labels of the representation
∇^2	Laplace operator
$\mathcal{G}$	Lie algebra
g	Lie algebra
G	Lie group
$\mathcal{V}$	Linear vector space
$\circ$	Long root
$\mathcal{F}$	Number field
O	Octonion
$\alpha^{(+)}$	Positive root
Q	Quaternion
$\langle\|\|\rangle$	Reduced matrix element
$\cdot$	Scalar product
$\oplus_s$	Semidirect sum
$\bullet$	Short root
SGA	Spectrum generating algebra
$\otimes$	Tensor product
$\oplus$	Tensor sum
Γ	Topological space
$|0\rangle$	Vacuum
$\mathcal{L}$	Vector space

Chapter 1
Basic Concepts

1.1 Definitions

The key notion in the definition of Lie algebras in physics is that of the commutator (or bracket), denoted by [,]. The commutator of X and Y is defined as

$$[X, Y] = XY - YX. \tag{1.1}$$

It satisfies the relations

$$[X, X] = 0; \qquad [X, Y] = -[Y, X]. \tag{1.2}$$

The operation [,] is in general neither commutative nor associative. The commutator $[X, Y]$ appears in quantum mechanics (Messiah 1958), and in classical mechanics, where it is called Poisson bracket (Goldstein et al. 2002) and often denoted by a curly bracket {, } instead of [,].

Another key notion is that of number field, $\mathcal{F}$. The number fields of interest are: Real, R, Complex, C, Quaternion, Q, and Octonion, O. Since these lecture notes are intended primarily for applications to quantal systems, where the basic commutation relations between coordinates and momenta are

$$[x, \frac{1}{i} \frac{\partial}{\partial x}] = i, \tag{1.3}$$

and to classical systems where the Poisson bracket is real, only real and complex fields will be considered. Although formulations of quantum mechanics in terms of quaternions and octonions have been suggested, Lie algebras over the quaternion and octonion number fields will not be discussed here.

© Springer-Verlag Berlin Heidelberg 2015
F. Iachello, *Lie Algebras and Applications*, Lecture Notes in Physics 891,
DOI 10.1007/978-3-662-44494-8_1

1.2   Lie Algebras

Lie algebras are named after the Norwegian mathematician Sophus Lie (1842–1899). Most of what we know about the original formulation comes from Lie's lecture notes in Leipzig, as collected by Scheffers (1893).

A set of elements $X_\alpha (\alpha = 1, \ldots, r)$ is said to form a Lie algebra $\mathcal{G}$, written as $X_\alpha \in \mathcal{G}$, if the following axioms are satisfied:

Axiom 1. The commutator of any two elements is a linear combination of the elements in the Lie algebra

$$[X_\rho, X_\sigma] = \sum_\tau c_{\rho\sigma}^\tau X_\tau. \tag{1.4}$$

Axiom 2. The double commutators of three elements satisfy the Jacobi identity

$$[X_\rho, [X_\sigma, X_\tau]] + [X_\sigma, [X_\tau, X_\rho]] + [X_\tau, [X_\rho, X_\sigma]] = 0. \tag{1.5}$$

The coefficients $c_{\rho\sigma}^\tau$ are called Lie structure constants. They define the Lie algebra. They satisfy

$$c_{\rho\sigma}^\tau = -c_{\sigma\rho}^\tau \tag{1.6}$$

and

$$c_{\rho\sigma}^\mu c_{\mu\tau}^\nu + c_{\sigma\tau}^\mu c_{\mu\rho}^\nu + c_{\tau\rho}^\mu c_{\mu\sigma}^\nu = 0. \tag{1.7}$$

A tensor notation with covariant, $X_{\rho\ldots}$, and contravariant, $X^{\rho\cdots}$, indices has been used in (1.4)–(1.7) and will be used in the remaining part of this chapter. In this notation, the structure constants $c_{\rho\sigma}^\tau$ are rank-3 tensors with one contravariant and two covariant indices. The outer product of two tensors, for example two covariant vectors, X_ρ and Y_σ, is written $X_\rho Y_\sigma$. The inner product of two tensors, for example two vectors, is written $X_\rho Y^\rho$. Also, unless otherwise specified, a summation convention over repeated indices will be used

$$c_{\rho\sigma}^\tau X_\tau \equiv \sum_\tau c_{\rho\sigma}^\tau X_\tau. \tag{1.8}$$

Algebras are usually denoted by script (calligraphic) letters, $\mathcal{G}$, or by lowercase letters, g. The associated Lie groups, to be discussed in Chap. 3, are usually denoted by capital letters, G. However, often no distinction is made between groups and algebras and the same letter is used for both. Unless otherwise specified, the notation lowercase-capital will be used in these notes.

The r elements, X_α, span a r-dimensional linear vector space $\mathcal{L}$. The definition given above in an 'operational' definition. A formal, abstract, 'mathematical' definition is:

Definition 1. A vector space $\mathcal{L}$ over a number field $\mathcal{F}$, with an operation $\mathcal{L} \times \mathcal{L} \to \mathcal{L}$, denoted $[X, Y]$ and called the commutator of X and Y, is called a Lie algebra over $\mathcal{F}$ if the following axioms are satisfied:

(1) The operation is bilinear.
(2) $[X, X] = 0$ for all X in $\mathcal{L}$.
(3) $[X, [Y, Z]] + [Y, [Z, X]] + [Z, [X, Y]] = 0$ $(X, Y, Z \in \mathcal{L})$.

The properties of bilinearity, i.e., $[aX + bY, Z] = a[X, Z] + b[Y, Z]$ and $[X, bY + cZ] = b[X, Y] + c[X, Z]$, and of vanishing $[X, X] = 0$ together guarantee the antisymmetry property (1.2) of the commutator. The closure of $\mathcal{L}$ under the commutator ($\mathcal{L} \times \mathcal{L} \to \mathcal{L}$) gives Axiom 1 above for the generators. Property (3) gives Axiom 2 above (Humphreys 1972).

A Lie algebra is called "real" if the field $\mathcal{F}$ is R, it is called "complex" if $\mathcal{F}$ is C. Real Lie algebras have real structure constants, while complex Lie algebras have structure constants which can be real or complex.

Example 1. The algebra

$$[X_1, X_2] = X_3 \ , \ \ [X_2, X_3] = X_1 \ , \ \ [X_3, X_1] = X_2 \tag{1.9}$$

is a real Lie algebra with three elements ($r = 3$).This is the angular momentum algebra in three dimensions, $so(3)$.

Example 2. The algebra

$$[X_1, X_2] = X_3 \ , \ \ [X_2, X_3] = -X_1 \ , \ \ [X_3, X_1] = X_2 \tag{1.10}$$

is also a real Lie algebra with three elements ($r = 3$). This is the Lorentz algebra in $2 + 1$ dimensions, $so(2, 1)$.

Note the difference between the two (a sign in the commutation relations).

1.3   Change of Basis

Let $X_\rho(\rho = 1, \ldots, r)$ be a basis in the r-dimensional vector space $\mathcal{L}$. It is possible to change the basis

$$X'_\sigma = a_\sigma^\rho \, X_\rho \tag{1.11}$$

where a_σ^ρ is non-singular. The new commutation relations of the algebra are

$$\left[X'_\rho, X'_\sigma\right] = c_{\rho\sigma}^{\prime\tau} X'_\tau. \tag{1.12}$$

From (1.4) and (1.11), the new structure constants $c'^\tau_{\rho\sigma}$ are obtained from the old structure constants $c^\kappa_{\nu\lambda}$ by solving the set of equations

$$c'^\tau_{\rho\sigma}\, a^\kappa_\tau = a^\nu_\rho\, a^\lambda_\sigma\, c^\kappa_{\nu\lambda}. \tag{1.13}$$

Particularly simple is the change of basis in which each element is multiplied by a real number (sometimes called a normalization transformation).

Example 3. The transformation

$$X'_1 = \sqrt{2}X_1 \ , \ \ X'_2 = \sqrt{2}X_2 \ , \ \ X'_3 = X_3 \tag{1.14}$$

changes the commutation relations of the Lie algebra $so(3)$ into

$$[X'_1, X'_2] = 2X'_3 \ ; \ \ [X'_2, X'_3] = X'_1 \ ; \ \ [X'_3, X'_1] = X'_2. \tag{1.15}$$

Lie algebras that have the same commutation relations up to a change of basis are called isomorphic. This definition is 'operational'. A formal, abstract, mathematical definition of isomorphism is:

Definition 2. Two Lie algebras g, g' over $\mathcal{F}$ are isomorphic if there exists a vector space isomorphism $\phi : g \rightarrow g'$ satisfying $\phi([X, Y]) = [\phi(X), \phi(Y)]$ for all X, Y in g. (Humphreys 1972, p.1) Isomorphism of algebras will be denoted by the symbol $\sim$. Isomorphisms of Lie algebras will be discussed in Chap. 2. An example is:

Example 4. The Lie algebras $so(3)$ and $su(2)$ are isomorphic

$$so(3) \sim su(2). \tag{1.16}$$

1.4   Complex Extensions

The change of basis (1.11) can be complex. An example is multiplication by the imaginary unit on $so(3)$,

$$J_1 = i\, X_1 \ , \ \ \ J_2 = i\, X_2 \ , \ \ \ J_3 = i\, X_3. \tag{1.17}$$

The commutation relations are now

$$[J_1, J_2] = i\, J_3 \ , \ [J_2, J_3] = i\, J_1 \ , \ [J_3, J_1] = i\, J_2. \tag{1.18}$$

The algebra composed of elements J_1, J_2, J_3 is the 'angular momentum algebra' often quoted in quantum mechanics books.

If one takes linear combinations of elements, A, B, of a real Lie algebra g, with complex coefficients, and defines $[A + iB, C] = [A, C] + i[B, C]$ one obtains the complex extension of the real Lie algebra. Starting from a real algebra g, by making a complex change of basis, one can construct a complex extension of g. In some cases, the complex extension of different real Lie algebras is the same.

Example 5. The real Lie algebras $so(2, 1)$ and $so(3)$ have the same complex extension.

Consider the real Lie algebra $so(2, 1)$ of Example 2. By making the change of basis

$$Y_1 = X_1 \quad , \quad Y_2 = -iX_2 \quad , \quad Y_3 = -iX_3 \quad , \tag{1.19}$$

one obtains

$$[Y_1, Y_2] = Y_3 \quad ; \quad [Y_2, Y_3] = Y_1 \quad ; \quad [Y_3, Y_1] = Y_2. \tag{1.20}$$

These are the same commutation relations of the real Lie algebra $so(3)$ of Example 1.

1.5 Lie Subalgebras

A subset of elements, Y_β, closed with respect to commutation is called a subalgebra

$$X_\alpha \in g; \qquad Y_\beta \in g'; \qquad g \supset g' \tag{1.21}$$

The symbol $\supset$ is used to indicate that g' is a subalgebra of g. The subset satisfies the commutation relations

$$\left[Y_\rho, Y_\sigma\right] = c_{\rho\sigma}^\tau Y_\tau \tag{1.22}$$

Example 6. The single element X_3 forms a Lie subalgebra of $so(3) \ni X_1, X_2, X_3$, since

$$[X_3, X_3] = 0. \tag{1.23}$$

This is $so(2)$, the angular momentum algebra in two dimensions, $so(3) \supset so(2)$.

1.6   Abelian Algebras

These are a special type of algebras, named after the Norwegian mathematician Niels Abel (1802–1829). An Abelian algebra, $\mathcal{A}$, is an algebra for which all elements commute,

$$[X_\rho, X_\sigma] = 0 \qquad \text{for any} \qquad X_\rho \in \mathcal{A} \ , \quad X_\sigma \in \mathcal{A}. \tag{1.24}$$

Example 7. The algebra $so(2) \ni X_3$ is Abelian, since

$$[X_3, X_3] = 0. \tag{1.25}$$

Any algebra contains a trivial Abelian subalgebra, composed of a single element X_ρ, since $\left[X_\rho, X_\rho\right] = 0$. Another non-trivial example is the translation algebra in two dimensions, $t(2) \ni X_1, X_2$.

Example 8. The algebra $t(2)$ with commutation relations

$$[X_1, X_2] = 0 \ , \ [X_1, X_1] = 0 \ , \ [X_2, X_2] = 0, \tag{1.26}$$

is Abelian.

1.7   Direct Sum

Consider two commuting algebras $g_1 \ni X_\alpha$, $g_2 \ni Y_\beta$, satisfying

$$[X_\rho, X_\sigma] = c_{\rho\sigma}^\tau X_\tau \qquad ,$$
$$[Y_\rho, Y_\sigma] = c_{\rho\sigma}^{\prime\tau} Y_\tau \qquad ,$$
$$[X_\rho, Y_\sigma] = 0. \tag{1.27}$$

The commuting property is denoted by $g_1 \cap g_2 = 0$. The set of elements X_α, Y_β forms an algebra g, called the direct sum,

$$g = g_1 \oplus g_2. \tag{1.28}$$

Sometimes, it is possible to rewrite a Lie algebra as a direct sum of other Lie algebras. Consider, the algebra $so(4) \ni X_1, X_2, X_3, Y_1, Y_2, Y_3$, satisfying commutation relations

$$[X_1, X_2] = X_3 \ ; \ [X_2, X_3] = X_1 \ ; \ [X_3, X_1] = X_2$$
$$[Y_1, Y_2] = X_3 \ ; \ [Y_2, Y_3] = X_1 \ ; \ [Y_3, Y_1] = X_2$$

$$[X_1, Y_1] = 0 \quad ; \quad [X_2, Y_2] = 0 \quad ; \quad [X_3, Y_3] = 0$$
$$[X_1, Y_2] = Y_3 \quad ; \quad [X_1, Y_3] = -Y_2$$
$$[X_2, Y_1] = -Y_3 \quad ; \quad [X_2, Y_3] = Y_1$$
$$[X_3, Y_1] = Y_2 \quad ; \quad [X_3, Y_2] = -Y_1 \tag{1.29}$$

By a change of basis

$$J_i = \frac{X_i + Y_i}{2} \quad , \quad K_i = \frac{X_i - Y_i}{2} \quad (i = 1, 2, 3) \tag{1.30}$$

the algebra can be brought to the form

$$[J_1, J_2] = J_3 \qquad [J_2, J_3] = J_1 \qquad [J_3, J_1] = J_2$$
$$[K_1, K_2] = K_3 \qquad [K_2, K_3] = K_1 \qquad [K_3, K_1] = K_2$$
$$[J_i, K_j] = 0 \quad (i, j = 1, 2, 3). \tag{1.31}$$

In the new form, one can see that the algebra $so(4)$ is the direct sum of two $so(3)$ algebras.

Example 9. The algebra $so(4)$ is isomorphic to the direct sums

$$so(4) \sim so(3) \oplus so(3) \sim su(2) \oplus su(2) \sim sp(2) \oplus sp(2). \tag{1.32}$$

The splitting is rarely possible. Consider for example, the algebra $so(3, 1) \ni X_1, X_2, X_3, Y_1, Y_2, Y_3$, satisfying commutation relations

$$[X_1, X_2] = X_3 \quad ; \quad [X_2, X_3] = X_1 \quad ; \quad [X_3, X_1] = X_2$$
$$[Y_1, Y_2] = -X_3 \quad ; \quad [Y_2, Y_3] = -X_1 \quad ; \quad [Y_3, Y_1] = -X_2$$
$$[X_1, Y_1] = 0 \quad ; \quad [X_2, Y_2] = 0 \quad ; \quad [X_3, Y_3] = 0$$
$$[X_1, Y_2] = Y_3 \quad ; \quad [X_1, Y_3] = -Y_2$$
$$[X_2, Y_1] = -Y_3 \quad ; \quad [X_2, Y_3] = Y_1$$
$$[X_3, Y_1] = Y_2 \quad ; \quad [X_3, Y_2] = -Y_1. \tag{1.33}$$

This algebra cannot be split into a direct sum of real Lie algebras. However, sometimes, the splitting is possible by going to the complex extension of the algebra. For example by taking the combination

$$J_j = \frac{X_j + iY_j}{2} \quad , \quad K_j = \frac{X_j - iY_j}{2} \quad (j = 1, 2, 3) \tag{1.34}$$

one can show that the elements J_j, K_j satisfy (1.31). The algebras $so(4)$ and $so(3,1)$ have the same complex extension, and can be split in the same fashion.

1.8   Ideals (Invariant Subalgebras)

Consider an algebra g and its subalgebra g', $X_\alpha \in g$, $Y_\beta \in g'$, $g \supset g'$. Since g' is a subalgebra, it satisfies

$$[Y_\rho, Y_\sigma] = c'^\tau_{\rho\sigma} Y_\tau \quad . \tag{1.35}$$

If, in addition,

$$[Y_\rho, X_\sigma] = c^\tau_{\rho\sigma} Y_\tau \quad , \tag{1.36}$$

then g' is called an invariant subalgebra (ideal) of g. As an example, consider the Euclidean algebra $e(2)$, composed of three elements, X_1, X_2, X_3, satisfying

$$[X_1, X_2] = X_3 \qquad [X_1, X_3] = -X_2 \qquad [X_2, X_3] = 0. \tag{1.37}$$

Example 10. $g' \ni X_2, X_3$ is an (Abelian) ideal of $g \equiv e(2) \ni X_1, X_2, X_3$.

1.9   Semisimple Algebras

An algebra which has no Abelian ideals is called semisimple.

Example 11. The algebra $so(3)$

$$[X_1, X_2] = X_3 \qquad [X_2, X_3] = X_1 \qquad [X_3, X_1] = X_2 \tag{1.38}$$

is semisimple.

Example 12. The algebra $e(2)$

$$[X_1, X_2] = X_3 \qquad [X_1, X_3] = -X_2 \qquad [X_2, X_3] = 0 \tag{1.39}$$

is non-semisimple.

Obviously g itself [and 0, the subspace consisting only of the zero vector] are ideals of g, called improper or trivial ideals. An algebra is called simple if it contains no ideals except g and 0. A simple Lie algebra is necessarily semisimple, though the converse need not hold. An additional condition for semisimplicity is that the algebra g not be Abelian, $[g, g] \neq 0$. This implies that simple and semisimple Lie algebras must necessarily contain more than one element. The algebras $so(2) \sim u(1)$ with commutation relation

$$[X_3, X_3] = 0 \tag{1.40}$$

cannot be classified as simple or semisimple, although $so(2)$ is often included in the classification of semisimple Lie algebras.

1.10   Semidirect Sum

If an algebra g has two subalgebras g_1, g_2 such that

$$[g_1, g_1] \in g_1, \quad [g_2, g_2] \in g_2, \quad [g_1, g_2] \in g_1, \tag{1.41}$$

then the algebra g is said to be the semidirect sum of g_1 and g_2. Clearly g_1 is an ideal of g. Note that g_1 does not to be an ideal of g_2. It suffices that it acts as in (1.41). Normally, one writes a semidirect sum by first giving the ideal and then the residual subalgebra, as

$$g = g_1 \oplus_s g_2. \tag{1.42}$$

Example 13. The Euclidean algebra $e(2)$, composed of three elements, X_1, X_2, X_3, is the semidirect sum of the rotation algebra in two dimensions, $so(2)$, composed of the single element, X_1, and the translation algebra in two dimensions, $t(2)$, composed of two commuting elements, X_2, X_3.

In the notation of (1.42)

$$e(2) = t(2) \oplus_s so(2). \tag{1.43}$$

1.11   Metric Tensor

With the Lie structure constants one can form a tensor, called metric tensor,

$$g_{\sigma\lambda} = g_{\lambda\sigma} = c_{\sigma\rho}^{\tau} c_{\lambda\tau}^{\rho}. \tag{1.44}$$

The metric tensor is also called Killing form named after Killing, who, in a series of papers in the 1880s, discussed its properties (Killing 1888,1889a,1889b,1890). The metric tensor is a geometric concept used by physicists. In the mathematical literature the Killing form is an algebraic concept (Humphreys 1972, p.21).

The metric tensor was used by Cartan to identify semisimple Lie algebras. Cartan's criterion for deciding if a Lie algebra is semisimple is:

Theorem 1. *A Lie algebra g is semisimple if, and only if,*

$$\det | g_{\sigma\lambda} | \neq 0 \ . \tag{1.45}$$

In other words, an inverse $g^{\sigma\lambda}$ of the metric tensor $g_{\sigma\lambda}$ exists

$$g^{\sigma\tau} g_{\tau\lambda} = \delta^{\sigma}_{\lambda}, \tag{1.46}$$

where

$$\delta^{\sigma}_{\lambda} = \begin{cases} 1 \text{ if } \sigma = \lambda \\ 0 \text{ if } \sigma \neq \lambda \end{cases} . \tag{1.47}$$

Example 14. The algebra $so(3)$ is semisimple.

The metric tensor of $so(3)$ is

$$g_{\sigma\lambda} = \begin{pmatrix} -2 & 0 & 0 \\ 0 & -2 & 0 \\ 0 & 0 & -2 \end{pmatrix}, \tag{1.48}$$

also written in compact form $g_{\sigma\lambda} = -2\delta_{\sigma\lambda}$. The determinant of the metric tensor is

$$\det | g_{\sigma\lambda} | = -8 \tag{1.49}$$

and thus the algebra $so(3)$ is semisimple.

Example 15. The algebra $so(2, 1)$ is semisimple.

The metric tensor of $so(2, 1)$ is

$$g_{\sigma\lambda} = \begin{pmatrix} -2 & 0 & 0 \\ 0 & +2 & 0 \\ 0 & 0 & +2 \end{pmatrix}. \tag{1.50}$$

The determinant of the metric tensor is again

$$\det | g_{\sigma\lambda} | = -8 \tag{1.51}$$

and thus the algebra $so(2, 1)$ is semisimple.

Example 16. The algebra $e(2)$ is non-semisimple.

Finally, consider the algebra $e(2)$ with metric tensor

$$g_{\sigma\lambda} = \begin{pmatrix} -2 & 0 & 0 \\ 0 & 0 & 0 \\ 0 & 0 & 0 \end{pmatrix}. \tag{1.52}$$

In this case

$$\det \mid g_{\sigma\lambda} \mid = 0 \tag{1.53}$$

and thus the algebra $e(2)$ is non-semisimple.

1.12 Compact and Non-Compact Algebras

A real semisimple Lie algebra is compact if its metric tensor is negative definite.

Example 17. The algebra $so(3)$ is compact.

The metric tensor of $so(3)$ is negative definite. In its diagonal form all elements are negative.

Example 18. The algebra $so(2, 1)$ is non-compact.

The metric tensor of $so(2, 1)$ is non-negative definite. In its diagonal form some elements are positive.

1.13 Derived Algebras

Starting with a Lie algebra, g, with elements X_ρ, it is possible to construct other algebras, called derived algebras, by taking commutators

$$g^{(0)} = g$$
$$g^{(1)} = [g^{(0)}, g^{(0)}]$$
$$g^{(2)} = [g^{(1)}, g^{(1)}]$$
$$\cdots \tag{1.54}$$

The sequence $g^{(0)}, g^{(1)}, g^{(2)}, \ldots, g^{(i)}$ is called a derived series. For example, starting with the Euclidean algebra, $e(2)$, with elements

$$g \doteq X_1, X_2, X_3 \tag{1.55}$$

satisfying the commutation relations (1.39), one has

$$g^{(1)} \doteq X_2, X_3$$
$$g^{(2)} \doteq 0. \tag{1.56}$$

If, for some positive k,

$$g^{(k)} \doteq 0 \tag{1.57}$$

the algebra is called solvable. The derived series should not be confused with derivation of g, denoted by $Der\ g$ [J.E. Humphreys, *loc.cit.*, p.4].

Example 19. The algebra $e(2)$ is solvable.

From (1.56),

$$[e(2)]^{(2)} \doteq 0. \tag{1.58}$$

1.14 Nilpotent Algebras

Starting with a Lie algebra, g, with elements, X_ρ, it is possible to construct another series, called descending central series, or lower central series, as

$$g^0 = g$$
$$g^1 = [g, g] = g^{(1)}$$
$$g^2 = [g, g^1]$$
$$\cdots$$
$$g^i = [g, g^{i-1}] \tag{1.59}$$

If, for some positive k,

$$g^k = 0 \tag{1.60}$$

the algebra is called nilpotent.

Example 20. The algebra $e(2)$ is not nilpotent

Starting with

$$g \doteq X_1, X_2, X_3, \tag{1.61}$$

satisfying (1.39), one has

$$g^1 \doteq X_2, X_3$$
$$g^2 \doteq X_2, X_3$$
$$\cdots \qquad (1.62)$$

1.15 Invariant Casimir Operators

These operators play a central role in applications. They are named after the Dutch physicist Casimir, who introduced them in 1931 for the angular momentum algebra $so(3)$ (Casimir 1931). An operator, C, that commutes with all the elements of a Lie algebra g

$$[C, X_\tau] = 0 \qquad \text{for all } X_\tau \in g \qquad (1.63)$$

is called an invariant Casimir operator. Casimir operators can be linear, quadratic, cubic, ... in the elements X_τ. They live in the enveloping algebra of g, defined in Sect. 2.16. A Casimir operator is called of order p if it contains products of p elements X_τ,

$$C_p = \sum_{\alpha_1...\alpha_p} f^{\alpha_1 \alpha_2 ... \alpha_p} X_{\alpha_1} X_{\alpha_2} ... X_{\alpha_p}. \qquad (1.64)$$

The summation is explicitly displayed in this formula. Also, if C commutes with g, so does aC, C^2,.... The number of independent Casimir operators of a Lie algebra will be discussed in Chap. 5. The quadratic ($p = 2$) Casimir operator of a semisimple algebra can be simply constructed from the metric tensor

$$C_2 = g^{\rho\sigma} X_\rho X_\sigma = g_{\rho\sigma} X^\rho X^\sigma \equiv C. \qquad (1.65)$$

Proof. Evaluate the commutator of C and X_τ

$$[C, X_\tau] = g^{\rho\sigma} [X_\rho X_\sigma, X_\tau] = g^{\rho\sigma} X_\rho [X_\sigma, X_\tau] + g^{\rho\sigma} [X_\rho, X_\tau] X_\sigma =$$
$$= g^{\rho\sigma} X_\rho c^\lambda_{\sigma\tau} X_\lambda + g^{\rho\sigma} c^\lambda_{\rho\tau} X_\lambda X_\sigma =$$
$$= g^{\rho\sigma} c^\lambda_{\sigma\tau} X_\rho X_\lambda + g^{\sigma\rho} c^\lambda_{\sigma\tau} X_\lambda X_\rho = g^{\rho\sigma} c^\lambda_{\sigma\tau} (X_\rho X_\lambda + X_\lambda X_\rho)$$
$$= g^{\rho\sigma} g^{\lambda\nu} c_{\nu\sigma\tau} (X_\rho X_\lambda + X_\lambda X_\rho) = 0 \qquad (1.66)$$

The last line follows from the fact that the product $g^{\sigma\lambda} g^{\lambda\nu}$ is symmetric under $\lambda \to \sigma$, $\nu \to \rho$, the product $(X_\rho X_\sigma + X_\sigma X_\rho)$ is symmetric under $\lambda \to \rho$, and the structure constant $c_{\nu\rho\sigma}$ is antisymmetric under $\nu \to \sigma$. For a semisimple Lie algebra, indices can be raised and lowered using the metric tensor

$$c^{\lambda}_{\sigma\tau} = g^{\lambda\nu} c_{\nu\sigma\tau}.$$ (1.67)

Higher order Casimir operators can be constructed in a similar fashion

$$C_p = c^{\beta_2}_{\alpha_1\beta_1} c^{\beta_3}_{\alpha_2\beta_2} \ldots c^{\beta_1}_{\alpha_p\beta_p} X^{\alpha_1} X^{\alpha_2} \ldots X^{\alpha_p}.$$ (1.68)

For the algebra $so(3)$, the inverse of the metric tensor is

$$g^{\sigma\lambda} = \begin{pmatrix} -\frac{1}{2} & 0 & 0 \\ 0 & -\frac{1}{2} & 0 \\ 0 & 0 & -\frac{1}{2} \end{pmatrix}$$ (1.69)

giving

$$C = -\frac{1}{2} (X_1^2 + X_2^2 + X_3^2).$$ (1.70)

For the algebra $so(2, 1)$

$$C = -\frac{1}{2} \left(X_1^2 - X_2^2 - X_3^2 \right).$$ (1.71)

Note the minus signs. By multiplying C by 2 and the elements by i, one obtains for $so(3)$

$$C' = 2C = J_1^2 + J_2^2 + J_3^2.$$ (1.72)

This is the usual form in which the Casimir operator of the angular momentum algebra $so(3)$ appears in quantum mechanics textbooks.

1.16   Invariant Operators for Non-Semisimple Algebras

For non-semisimple Lie algebras, Casimir operators cannot be simply constructed. One introduces a related notion of invariant operators that commute will all elements.

Example 21. The invariant operator of the Euclidean algebra $e(2) \ni X_1, X_2, X_3$ is

$$C = X_2^2 + X_3^2$$ (1.73)

Proof. The commutators are

$$\left[X_2^2, X_1\right] = X_2[X_2, X_1] + [X_2, X_1] \, X_2 = -2 \, X_2 \, X_3$$
$$\left[X_3^2, X_1\right] = X_3[X_3, X_1] + [X_3, X_1] \, X_3 = 2 \, X_2 \, X_3. \tag{1.74}$$

Hence

$$[C, X_\tau] = 0 \text{ for any } X_\tau \in g. \tag{1.75}$$

1.17 Contractions of Lie Algebras

Let $X_1, \ldots, X_r$ be the elements of the Lie algebra g. For a subset $X_1, \ldots, X_\rho, \rho \leq r$ define

$$Y_i = \varepsilon X_i, \quad i = 1, \ldots, \rho \leq r, \tag{1.76}$$

and express the commutation relations in terms of the Y_i,

$$[Y_i, Y_j] = c_{ij}^k \varepsilon Y_k + c_{ij}^m \varepsilon^2 X_m,$$
$$[Y_i, X_m] = c_{im}^k Y_k + c_{im}^n \varepsilon X_n,$$
$$[X_m, X_n] = c_{mn}^i \varepsilon^{-1} Y_i + c_{mn}^s X_s,$$
$$i, j, k \leq \rho, \quad \rho < m, n, s \leq r. \tag{1.77}$$

Now, let $\varepsilon \to 0$. If

$$c_{mn}^i = 0, \quad i \leq \rho, \quad \rho < m, n \leq r, \tag{1.78}$$

the commutation relations

$$[Y_i, Y_j] = 0, \quad [Y_i, X_k] = c_{im}^k Y_k, \quad [X_m, X_s] = c_{mn}^s X_s, \tag{1.79}$$

define a Lie algebra, called the Inonu contracted Lie algebra g'

$$g \longrightarrow_c g'. \tag{1.80}$$

Example 22. The Euclidean algebra $e(2)$ is a contraction of the Lorentz algebra $so(2, 1)$.

The commutation relations of the complex extension of $so(2, 1)$ are

$$[J_1, J_2] = i J_3, \quad [J_2, J_3] = -i J_1, \quad [J_3, J_1] = i J_2, \tag{1.81}$$

with Casimir operator

$$C = J_1^2 - J_2^2 - J_3^2. \tag{1.82}$$

Introducing

$$P_x = \varepsilon J_2, \quad P_y = \varepsilon J_3, \quad L_z = J_1, \tag{1.83}$$

the commutation relations become

$$\left[P_x, P_y\right] = -i\varepsilon^2 L_z, \quad \left[L_z, P_y\right] = -iP_x, \quad \left[L_z, P_x\right] = iP_y. \tag{1.84}$$

Letting $\varepsilon \to 0$, we obtain the commutation relations (1.39) of $e(2)$,

$$\left[P_x, P_y\right] = 0, \quad \left[L_z, P_y\right] = -iP_x, \quad \left[L_z, P_x\right] = iP_y, \tag{1.85}$$

with invariant operator

$$C' = P_x^2 + P_y^2. \tag{1.86}$$

[Also $so(3)$ contracts to $e(2)$ since $so(3)$ and $so(2, 1)$ are complex extensions of each other].

Example 23. The Euclidean algebra $e(3)$ is a contraction of the Lorentz algebra $so(3, 1)$.

The commutation relations of $so(3, 1)$ written in terms of elements L_i, K_i, ($i = 1, 2, 3$), are

$$\left[L_i, L_j\right] = i\,\epsilon_{ijk}L_k, \quad \left[L_i, K_j\right] = i\,\epsilon_{ijk}K_k, \quad \left[K_i, K_j\right] = -i\,\epsilon_{ijk}L_k, \tag{1.87}$$

where ϵ_{ijk} is the Levi-Civita symbol. Defining $P_i = \lambda^{-1}K_i (i = 1, 2, 3)$ and letting $\lambda \to \infty$, they become,

$$\left[L_i, L_j\right] = i\,\epsilon_{ijk}L_k, \quad \left[L_i, P_j\right] = i\,\epsilon_{ijk}P_k, \quad \left[P_i, P_j\right] = 0. \tag{1.88}$$

(Here $\lambda = \varepsilon^{-1}$ is used not to confuse it with the Levi-Civita symbol). These are the commutation relations of the algebra $e(3)$ composed of the three components P_1, P_2, P_3 of the momentum and the three components L_1, L_2, L_3 of the angular momentum in three dimensions. The algebra $e(3)$ is the semidirect sum of $so(3)$ and $t(3)$,

$$e(3) = t(3) \oplus_s so(3). \tag{1.89}$$

Example 24. The Poincare' algebra $p(4)$ is a contraction of the de Sitter algebra $so(3, 2)$.

The commutation relations of $so(3,2)$ written in terms of the 10 elements $M_{ab} = -M_{ba}$ $(a,b = 1,\ldots,5)$ are

$$[M_{ab}, M_{cd}] = -(g_{bc}M_{ad} - g_{ac}M_{bd} + g_{ad}M_{bc} - g_{bd}M_{ac}), \tag{1.90}$$

with $g_{ab} = (-,-,-,+,+)$. Defining $P_\mu = \varepsilon M_{5\mu}$, and letting $\varepsilon \to 0$, we obtain the commutation relations of $p(4)$, the Poincare' algebra in four dimensions,

$$[P_\mu, P_\nu] = 0, \quad [M_{\mu\nu}, P_\mu] = -g_{\nu\rho}P_\mu - g_{\mu\rho}P_\nu, \quad \mu,\nu,\rho = 1,2,3,4. \tag{1.91}$$

The algebra $p(4)$ composed of the four components, P_μ, of the four-momentum and the six components, $M_{\mu\nu}$, of the angular momentum and boost, is the semidirect sum of $so(3,1)$ and $t(3,1)$

$$p(4) = t(3,1) \oplus_s so(3,1). \tag{1.92}$$

[Also, $so(4,1)$, with metric $g_{ab} = (-,-,-,+,-)$, contracts to $p(4)$].

The relevance of the contraction process is that the representation theory of the contracted Lie algebras g' can then be done starting from that of the Lie algebras g and letting $\varepsilon \to 0$. Therefore, the contraction process greatly simplifies the treatment of Euclidean and Poincare' algebras.

1.18 Structure of Lie Algebras

The structure of Lie algebras, semisimple or not, can be investigated by inspection. A detailed account is given by Kirillov (1976).

1.18.1 Algebras with One Element

We begin with the case $r = 1$. In this case there is only one element, X, and one possibility

$$(a) \quad [X, X] = 0 \ . \tag{1.93}$$

The algebra is Abelian.

Example 22. The algebras $so(2) \sim u(1)$ are examples of Kirillov's case 1a.

1.18.2 Algebras with Two Elements

Next consider the case $r = 2$. In this case, there are two elements, X_1, X_2, and two possibilities

$$(a) \quad [X_1, X_2] = 0, \tag{1.94}$$

and

$$(b) \quad [X_1, X_2] = X_1 \quad . \tag{1.95}$$

In case (a), the algebra is Abelian. In case (b), X_1 is an Abelian ideal.

Example 23. The translation algebra $t(2)$ is an example of Kirillov's case 2a.

1.18.3 Algebras with Three Elements

For $r = 3$, there are three elements, X_1, X_2, X_3 and four possibilities:

$$(a) \quad [X_1, X_2] = [X_2, X_3] = [X_3, X_1] = 0 \tag{1.96}$$

$$(b) \quad [X_1, X_2] = X_3; \quad [X_1, X_3] = [X_2, X_3] = 0 \quad \text{or}$$
$$[X_1, X_3] = X_2; \quad [X_1, X_2] = [X_2, X_3] = 0, \tag{1.97}$$

$$(c) \quad [X_1, X_2] = 0; \ [X_3, X_1] = \alpha X_1 + \beta X_2; \ [X_3, X_2] = \gamma X_1 + \delta X_2, \tag{1.98}$$

where the matrix $\begin{vmatrix} \alpha & \beta \\ \gamma & \delta \end{vmatrix}$ is non-singular, and

$$(d) \quad [X_1, X_2] = X_3; \quad [X_2, X_3] = X_1; \quad [X_3, X_1] = X_2 \qquad \text{or}$$
$$[X_1, X_2] = X_3; \quad [X_2, X_3] = -X_1; \quad [X_3, X_1] = X_2 \quad . \tag{1.99}$$

In case (a), the algebra is Abelian.

Example 24. The translation algebra in three dimensions $t(3)$ is an example of Kirillov's case 3a.

Example 25. The Euclidean algebra $e(2)$ is Kirillov's case 3c, with $\alpha = 0, \beta = 1, \gamma = -1, \delta = 0$.

Example 26. The algebras $so(3)$ *and* $so(2, 1)$ are examples of Kirillov's case 3d.

This procedure becomes very cumbersome as the number of elements in the algebra increases. However, it allows one to classify all Lie algebras, including some non-semisimple algebras of physical interest, such as the Euclidean algebras. It has explicitly been carried out to $r = 8$. A different method, due to Cartan, provides a classification of semisimple Lie algebras for any number of elements.

Further details on the basic concepts of Lie algebras, including the somewhat elaborate constructions and proofs of Lie theory, can be found in (Bröcker and tom Dieck 1985); (Chaichian and Hagedorn 1998); (Cornwell 1997); (Duistermaat and Kolk 2000); (Fulton and Harris 1991); (Hall 2003); (Hermann 1966); (Hladik 1999); (Jacobson 1962); (Kirillov 2008); (Sattinger and Weaver 1986); (Serre 1965); (Weyl 1925).

Chapter 2
Semisimple Lie Algebras

2.1 Cartan–Weyl Form of a (Complex) Semisimple Lie Algebra

In 1894, the French mathematician Cartan provided a way to classify all semisimple Lie algebras (Cartan 1894). The subject was subsequently taken up by Weyl (1926) and van der Waerden (1933). We begin by rewriting the algebra as

$$X_\sigma \equiv (H_i, E_\alpha) \; ; \; (i = 1, \dots, l). \tag{2.1}$$

The elements H_i form the maximal Abelian subalgebra, often called the Cartan subalgebra,

$$[H_i, H_k] = 0 \qquad (i, k = 1, \dots, l). \tag{2.2}$$

The number of elements in the Cartan subalgebra, l, is called the rank of the algebra. The commutation relations of H_i with E_α are

$$[H_i, E_\alpha] = \alpha_i \, E_\alpha \tag{2.3}$$

while those of the E's among themselves are

$$[E_\alpha, E_\beta] = N_{\alpha\beta} \, E_{\alpha+\beta} \qquad (\text{if } \alpha + \beta \neq 0) \tag{2.4}$$

$$[E_\alpha, E_{-\alpha}] = \alpha^i \, H_i. \tag{2.5}$$

The α_i's are called roots and $N_{\alpha\beta}$ is a normalization. This form of the Lie algebra is called the Cartan–Weyl form.

© Springer-Verlag Berlin Heidelberg 2015
F. Iachello, *Lie Algebras and Applications*, Lecture Notes in Physics 891,
DOI 10.1007/978-3-662-44494-8_2

2.2 Graphical Representation of Root Vectors

One considers the α_i's ($i = 1, \ldots, l$) as the components of a covariant vector lying in an l-dimensional weight space with scalar product

$$(\alpha, \beta) \equiv \alpha^i \beta_i = \alpha_i \beta^i. \tag{2.6}$$

Van der Waerden derived a set of rules for the algebra to be a semisimple Lie algebra.

Rule 1. *If α is a root, so is $-\alpha$.*

Rule 2. *If α, β are roots, $\frac{2(\alpha, \beta)}{(\alpha, \alpha)}$ is an integer.*

Rule 3. *If α, β are roots, $\beta - 2\alpha \frac{(\alpha, \beta)}{(\alpha, \alpha)}$ is a root.*

From these, it follows that the angle φ between roots

$$\cos \varphi = \frac{(\alpha, \alpha)}{\sqrt{(\alpha, \alpha)(\beta, \beta)}} \tag{2.7}$$

can take the values

$$\cos^2 \varphi = 0, \frac{1}{4}, \frac{1}{2}, \frac{3}{4}, 1$$

$$\varphi = 0°, 30°, 45°, 60°, 90° \tag{2.8}$$

Roots can be displayed graphically in a root diagram.

For rank $l = 1$, the root diagram is a line, and there is only one possibility (Fig. 2.1).

The algebra, called A_1 by Cartan and $so(3) \sim su(2)$ by physicists, has three elements, $r = 3$.

For rank $l = 2$, the root diagram is planar. There are several possibilities (Fig. 2.2):

1. $\varphi = 30°$

    This algebra called G_2 by Cartan, has 14 elements, $r = 14$.
2. $\varphi = 45°$

    This algebra called B_2 by Cartan and $so(5)$ by physicists has 10 elements, $r = 10$.
3. $\varphi = 60°$

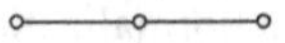

$$so(3)\sim su(2)\sim A_1$$

Fig. 2.1 Root diagram of the rank $l = 1$ algebra

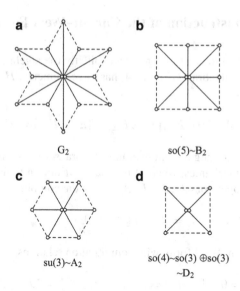

a G_2

b $so(5)\sim B_2$

c $su(3)\sim A_2$

d $so(4)\sim so(3) \oplus so(3)$
$\sim D_2$

Fig. 2.2 Root diagrams of rank $l = 2$ algebras

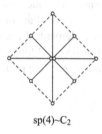

$sp(4)\sim C_2$

Fig. 2.3 Root diagram of the algebra C_2

This algebra, called A_2 by Cartan and $su(3)$ by physicists has 8 elements, $r = 8$.

4. $\varphi = 90°$

This algebra, called D_2 by Cartan and $so(4)$ by physicists, has $r = 6$ elements. It can be seen as the direct sum of two A_1 algebras, $D_2 \sim A_1 \oplus A_1$ or $so(4) \sim so(3) \oplus so(3)$.

The Cartan classification of rank-2 algebras contains also the algebra C_2, called $sp(4)$ by physicists. The root diagram of C_2 is identical to that of B_2 rotated by 45° (Fig. 2.3).

The algebras C_2 and B_2 are isomorphic, $sp(4) \sim so(5)$.

For rank $l = 3$ the root diagram is three-dimensional and it will not be drawn here.

2.3   Explicit Construction of the Cartan–Weyl Form

For applications, it is of interest to construct explicitly Lie algebras in Cartan–Weyl form. The rank $l = 1$ algebra, A_1 (Fig. 2.4) has three elements, H_1, E_{+1}, E_{-1}, with commutation relations,

$$[H_1, H_1] = 0; \quad [H_1, E_{\pm 1}] = \pm E_{\pm 1}; \quad [E_{+1}, E_{-1}] = H_1 \tag{2.9}$$

This algebra, being the angular momentum algebra $su(2) \sim so(3)$, is of great interest in quantum mechanics, where it is usually written in terms of the physical angular momentum operators J_x, J_y, J_z. By taking the combinations

$$J_z = J_z \; , \; J_\pm = J_x \pm i J_y \tag{2.10}$$

one obtains the Cartan–Weyl form, with commutation relations

$$[J_z, J_z] = 0; \quad [J_z, J_\pm] = \pm J_\pm; \quad [J_+, J_-] = 2 J_z. \tag{2.11}$$

The factor of two in the last commutator is due to a different normalization of the elements of the algebra. One can also see that, in this case, the Abelian Cartan subalgebra is composed of only one element, J_z, while the Weyl elements are J_+ and J_-. These elements are called raising and lowering operators.

Another important construction is that of the rank $l = 2$ algebra A_2 (Fig. 2.5). This algebra has 8 elements,

$$H_1, H_2 \quad , \quad E_{\pm \alpha} \quad , \quad E_{\pm \beta} \quad , \quad E_{\pm(\alpha+\beta)} \tag{2.12}$$

with commutation relations

$$[H_1, E_{\pm \alpha}] = \pm \frac{1}{2\sqrt{3}} E_{\pm \alpha} \qquad [H_2, E_{\pm \alpha}] = \pm \frac{1}{2} E_{\pm \alpha}$$

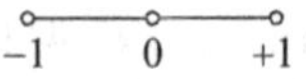

```
    o-------o-------o
   -1       0      +1
```

Fig. 2.4 Root diagram of the algebra A_1 with roots explicitly displayed

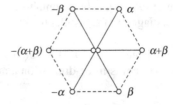

Fig. 2.5 Root diagram of the algebra A_2 with roots explicitly displayed

$$[H_1, E_{\pm\beta}] = \pm\frac{1}{2\sqrt{3}}E_{\pm\beta} \qquad [H_2, E_{\pm\beta}] = \pm\frac{1}{2}E_{\pm\beta}$$

$$[H_1, E_{\pm(\alpha+\beta)}] = \pm\frac{1}{\sqrt{3}}E_{\pm(\alpha+\beta)} \qquad [H_2, E_{\pm(\alpha+\beta)}] = 0$$

$$[E_\alpha, E_{-\alpha}] = \frac{H_1}{2\sqrt{3}} + \frac{H_2}{2} \qquad [E_\beta, E_{-\beta}] = \frac{H_1}{2\sqrt{3}} - \frac{H_2}{2}$$

$$[E_{(\alpha+\beta)}, E_{-(\alpha+\beta)}] = \frac{H_1}{\sqrt{3}} \qquad [E_\alpha, E_\beta] = \frac{1}{\sqrt{6}}E_{\alpha+\beta}$$

$$[E_\alpha, E_{(\alpha+\beta)}] = 0 \qquad [E_\beta, E_{(\alpha+\beta)}] = 0$$

$$[E_\alpha, E_{-(\alpha+\beta)}] = -\frac{1}{\sqrt{6}}E_{-\beta} \qquad [E_\beta, E_{-(\alpha+\beta)}] = \frac{1}{\sqrt{6}}E_{-\alpha}$$

$$[H_i, H_j] = 0 \quad (i, j = 1, 2) \tag{2.13}$$

The algebra has two Cartan elements and six Weyl elements (raising and lowering operators).

2.4 Dynkin Diagrams

The root diagrams for rank $l \geq 3$ cannot be displayed easily. The Russian mathematician Dynkin devised a method to display root diagrams of all semisimple Lie algebras (Dynkin 1947, 1962). We begin by introducing the notion of positive root.

Definition 1. Positive roots, $\alpha^{(+)}$, are those for which, in some arbitrary frame, its first coordinate different from zero is positive.

The number of positive roots is half of non-null roots.

Example 1. Root diagram of B_2

The roots are $(1,0)$ $(1,1)$ $(0,1)$ $(-1,1)$ $(-1,0)$ $(-1,-1)$ $(0,-1)$ $(1,-1)$. The positive roots are $(1,0)$ $(1,1)$ $(0,1)$ $(1,-1)$, with sum $\sum \alpha^{(+)} = (3,1)$. We next introduce the notion of simple roots.

Definition 2. A simple root is a positive root which cannot be decomposed into the sum of positive roots.

In the case of B_2 (Fig. 2.6), the two roots $(1,0)$ and $(1,1)$ can be decomposed as $(1,0) = (1,-1) + (0,1)$ and $(1,1) = (1,0) + (0,1)$. The simple roots are thus only $\alpha \equiv (0,1)$ and $\beta \equiv (1,-1)$. Dynkin showed that the angle between two simple roots can only be $90°, 120°, 135°, 150°$ and the normalization

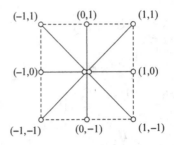

Fig. 2.6 Root diagram of the algebra B_2 with roots explicitly displayed

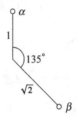

Fig. 2.7 Simple roots of B_2

$$\frac{(\beta,\beta)}{(\alpha,\alpha)} = \begin{cases} 1 & \vartheta_{\alpha,\beta} = 120° \\ 2 & \vartheta_{\alpha,\beta} = 135° \\ 3 & \vartheta_{\alpha,\beta} = 150° \\ \text{undetermined} & \vartheta_{\alpha,\beta} = 90° \end{cases} \qquad (2.14)$$

The root α is called short and the root β is called long. When plotted as before, the simple root diagram of B_2 appears as in Fig. 2.7. All information on the Lie algebra can then be condensed into a Dynkin diagram. In this diagram, the angle ϑ between roots is indicated by

$$\vartheta = 120° \quad \text{single line}$$
$$\vartheta = 135° \quad \text{double line}$$
$$\vartheta = 150° \quad \text{triple line}$$
$$\vartheta = 90° \quad \text{not joined} \qquad (2.15)$$

A short root is indicated by a filled dot $\bullet$, while a long root is indicated by an open dot $\circ$.

Example 2. Dynkin diagrams of rank two algebras

Dynkin diagrams of rank two algebras are shown in Fig. 2.8.

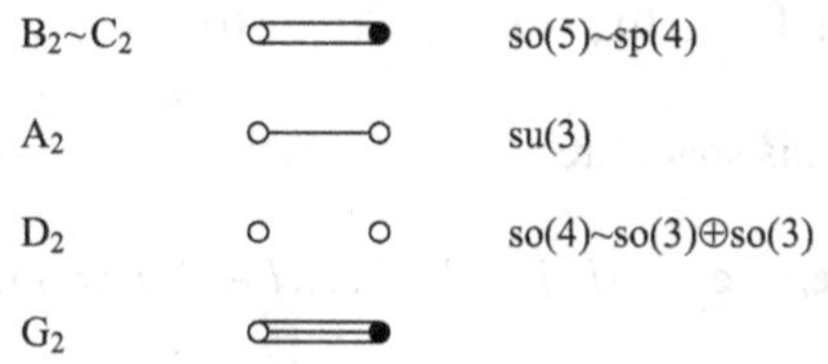

$B_2 \sim C_2$	o⬤	so(5)~sp(4)
A_2	o—o	su(3)
D_2	o o	so(4)~so(3)⊕so(3)
G_2	o⬤	

Fig. 2.8 Dynkin diagrams of rank-2 algebras

Table 2.1 Cartan classification of complex semisimple Lie algebras (and $u(n)$)

Name	Label		Cartan	Order (r)	Rank (l)
Special unitary	$su(n)$		A_l	$n^2 - 1$	$n - 1$
[Special] orthogonal	$so(n)$	(n odd)	B_l	$n(n-1)/2$	$(n-1)/2$
Symplectic	$sp(n)$	(n even)	C_l	$n(n+1)/2$	$n/2$
[Special] orthogonal	$so(n)$	(n even)	D_l	$n(n-1)/2$	$n/2$
Exceptional	G_2		G_2	14	2
	F_4		F_4	52	4
	E_6		E_6	78	6
	E_7		E_7	133	7
	E_8		E_8	248	8
Unitary	$u(n)$		–	n^2	n

2.5 Classification of (Complex) Semisimple Lie Algebras

All complex semisimple Lie algebras have been classified by Cartan and are given in Table 2.1. Because of its importance, the non-semisimple Lie algebra $u(n)$ is included as well. This algebra is of order n^2 and rank n.

There is no difference between orthogonal and special orthogonal algebras and hence 'special' has been put in brackets in the table. The Abelian algebra $so(2) \sim u(1)$ composed of a single element is included in the Cartan classification under $so(n)$ n =even, although strictly speaking it is not possible to apply to it Cartan's criterion since it does not have any subalgebra except itself.

2.6 Rules for Constructing the Root Vector Diagrams of Classical Lie Algebras

Van der Waerden derived a set of rules for constructing the root-vector diagrams of all complex semisimple Lie algebras. The rules for the classical Lie algebras A_l, B_l, C_l, and D_l are:

1. $A_l \sim su(l+1)$. Introduce $l+1$ mutually orthogonal vectors $\mathbf{e}_i$ ($i = 1, 2, \ldots, l+1$) in a $(l + 1)$-dimensional space

$$\mathbf{e}_1 = (1, 0, \ldots, 0); \quad \mathbf{e}_2 = (0, 1, \ldots, 0); \quad \ldots \quad . \tag{2.16}$$

The root vectors in this space are

$$\mathbf{e}_i - \mathbf{e}_j \quad (i, j = 1, 2, \ldots, l + 1; i \neq j). \tag{2.17}$$

Project these vectors into the hyperplane orthogonal to the vector $\mathbf{e}_1 + \ldots + \mathbf{e}_{l+1}$.

2. $B_l \sim so(2l + 1)$. Introduce l mutually orthogonal unit vectors $\mathbf{e}_i$ ($i = 1, \ldots, l$) in an l-dimensional space. The root vectors are

$$\pm \mathbf{e}_i, \quad \pm(\mathbf{e}_i \pm \mathbf{e}_j); \quad i \neq j \tag{2.18}$$

3. $C_l = sp(2l)$. Same as B_l but with root vectors

$$\pm 2\mathbf{e}_i, \quad \pm(\mathbf{e}_i \pm \mathbf{e}_j); \quad i \neq j \tag{2.19}$$

4. $D_l = so(2l)$. Same as B_l but with root vectors

$$\pm(\mathbf{e}_i \pm \mathbf{e}_j); \quad i \neq j \tag{2.20}$$

Example 3. Root vector diagram of $A_1 \sim su(2)$

There are two mutually orthogonal vectors in the $l + 1$ space

$$\mathbf{e}_1 = (1, 0), \quad \mathbf{e}_2 = (0, 1). \tag{2.21}$$

The root vectors are

$$\pm(\mathbf{e}_1 - \mathbf{e}_2) \tag{2.22}$$

with coordinates $(1, -1), (-1, 1)$. There are two roots plus one null root for a total of 3 roots. The root vector diagram is shown in Fig. 2.9. This figure is the same as Fig. 2.4 rotated by 45°.

Example 4. Root vector diagram of $B_2 \sim so(5)$

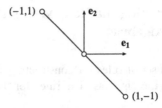

Fig. 2.9 Root vector diagram of $su(2)$

For B_2, there are two orthogonal unit vectors

$$\mathbf{e}_1 = (1,0), \quad \mathbf{e}_2 = (0,1). \tag{2.23}$$

The root vector are

$$\pm \mathbf{e}_1, \quad \pm \mathbf{e}_2, \quad \pm \mathbf{e}_1 \pm \mathbf{e}_2. \tag{2.24}$$

The coordinates of the end points are $(\pm 1, 0)$, $(0, \pm 1)$, $(1, \pm 1)$, $(-1, \pm 1)$. There are 8 roots plus two null roots for a total of 10 roots. The root vector diagram is shown in Fig. 2.6.

2.7 Rules for Constructing Root Vector Diagrams of Exceptional Lie Algebras

Rules for constructing the root vector diagrams of exceptional Lie algebras are rather complicated (Humphreys 1972).

1. G_2. Add to the 6 non-null roots of $A_2 \sim su(3)$

$$\mathbf{e}_i - \mathbf{e}_j \quad (i, j = 1, 2, 3; i \neq j) \tag{2.25}$$

the 6 root vectors

$$\pm \left(2\mathbf{e}_i - \mathbf{e}_j - \mathbf{e}_k\right) \quad (i, j, k = 1, 2, 3; i \neq j \neq k). \tag{2.26}$$

These 12 roots plus the 2 null roots give the 14 roots of G_2.

2. F_4. Add to the 32 non-null roots of $B_4 \sim so(9)$

$$\pm \mathbf{e}_i, \quad \pm (\mathbf{e}_i \pm \mathbf{e}_j); \quad i \neq j \quad (i, j = 1, 2, 3, 4) \tag{2.27}$$

the 16 root vectors

$$\frac{1}{2} (\pm \mathbf{e}_1, \pm \mathbf{e}_2 \pm \mathbf{e}_3 \pm \mathbf{e}_4) . \tag{2.28}$$

These 48 roots plus the 4 null roots give the 52 roots of F_4.

3. E_6, E_7, E_8. Rules for constructing the root systems of these algebras are very complicated and will not be given here. For E_8 they are given in Humphreys (1972), p.65. Those of E_6 and E_7 can be obtained from those of E_8 by inclusion of the Dynkin diagram of E_6 into E_7 and of E_7 into E_8.

2.8   Dynkin Diagrams of Classical Lie Algebras

The Dynkin diagrams of all classical Lie algebras can be obtained from the rules of
the previous Sects. 2.4 and 2.6 and are given in Fig. 2.10. In this figure, the simple
roots are denoted by $\alpha_1, \alpha_2, \ldots, \alpha_l$.

2.9   Dynkin Diagrams of Exceptional Lie Algebras

The Dynkin diagrams of exceptional Lie algebras are given in Fig. 2.11.

2.10   Cartan Matrices

Another way of condensing the information on a given Lie algebra is by constructing
Cartan matrices. Let $\Pi \equiv (\alpha_1, \ldots, \alpha_l)$ be a system of simple roots. The matrices

$$A_{ij} = \frac{2\left(\alpha_i, \alpha_j\right)}{\left(\alpha_i, \alpha_i\right)} \tag{2.29}$$

are called Cartan matrices. The diagonal elements are $A_{ii} = 2$. The off-diagonal
elements can be obtained from

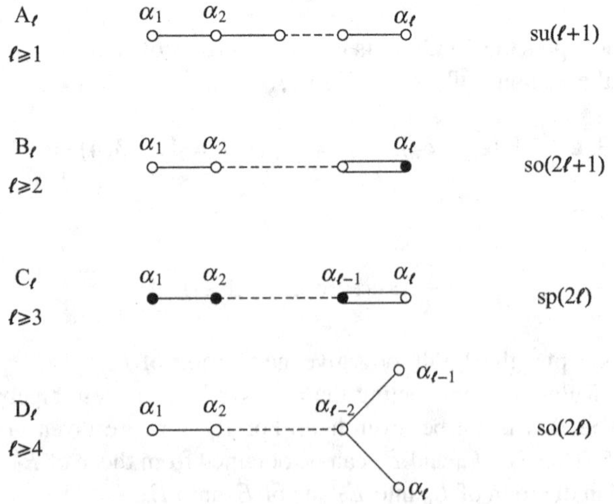

Fig. 2.10 Dynkin diagrams of classical Lie algebras

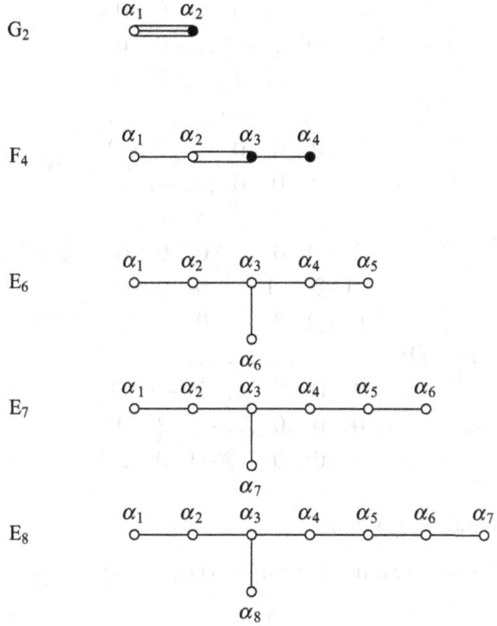

$$G_2 \quad \overset{\alpha_1 \quad \alpha_2}{\Longleftarrow}$$

$$F_4 \quad \overset{\alpha_1 \quad \alpha_2 \quad \alpha_3 \quad \alpha_4}{\circ - \circ \Longrightarrow \bullet - \bullet}$$

$$E_6 \quad \overset{\alpha_1 \quad \alpha_2 \quad \alpha_3 \quad \alpha_4 \quad \alpha_5}{\circ - \circ - \circ - \circ - \circ}$$
$$\underset{\alpha_6}{|}$$

$$E_7 \quad \overset{\alpha_1 \quad \alpha_2 \quad \alpha_3 \quad \alpha_4 \quad \alpha_5 \quad \alpha_6}{\circ - \circ - \circ - \circ - \circ - \circ}$$
$$\underset{\alpha_7}{|}$$

$$E_8 \quad \overset{\alpha_1 \quad \alpha_2 \quad \alpha_3 \quad \alpha_4 \quad \alpha_5 \quad \alpha_6 \quad \alpha_7}{\circ - \circ - \circ - \circ - \circ - \circ - \circ}$$
$$\underset{\alpha_8}{|}$$

Fig. 2.11 Dynkin diagrams of exceptional Lie algebras

Theorem 1. *If $\alpha, \beta \in \Pi$, then $\frac{2(\alpha,\beta)}{(\alpha,\alpha)} = -p$, where p is a positive integer. Since $\frac{(\beta,\beta)}{(\alpha,\alpha)}$ is given by (2.14), the off-diagonal elements are restricted to $0, -1, -2, -3$.*

2.11 Cartan Matrices of Classical Lie Algebras

The Cartan matrices of the classical Lie algebras constructed from the Dynkin diagrams of Sect. 2.8 are given by

$$A_l : \begin{vmatrix} 2 & -1 & 0 & \dots & 0 & 0 \\ -1 & 2 & -1 & \dots & 0 & 0 \\ 0 & -1 & 2 & \dots & 0 & 0 \\ \dots & \dots & \dots & \dots & \dots & \dots \\ 0 & 0 & 0 & \dots & 2 & -1 \\ 0 & 0 & 0 & \dots & -1 & 2 \end{vmatrix} \qquad (2.30)$$

$$B_l, C_l : \begin{vmatrix} 2 & -1 & 0 & \dots & 0 & 0 \\ -1 & 2 & -1 & \dots & 0 & 0 \\ 0 & -1 & 2 & \dots & 0 & 0 \\ \dots & \dots & \dots & \dots & \dots & \dots \\ 0 & 0 & 0 & \dots & 2 & -2 \\ 0 & 0 & 0 & \dots & -1 & 2 \end{vmatrix} \qquad (2.31)$$

$$D_l : \begin{vmatrix} 2 & -1 & 0 & \dots & 0 & 0 & 0 \\ -1 & 2 & -1 & \dots & 0 & 0 & 0 \\ 0 & -1 & 2 & \dots & 0 & 0 & 0 \\ \dots & \dots & \dots & \dots & \dots & \dots & \dots \\ 0 & 0 & 0 & \dots & 2 & -1 & -1 \\ 0 & 0 & 0 & \dots & -1 & 2 & 0 \\ 0 & 0 & 0 & \dots & -1 & 0 & 2 \end{vmatrix}. \qquad (2.32)$$

Example 5. Cartan matrix of $su(3)$

For the algebra $A_2 \sim su(3)$, the Cartan matrix is

$$su(3) : \begin{vmatrix} 2 & -1 \\ -1 & 2 \end{vmatrix}. \qquad (2.33)$$

2.12   Cartan Matrices of Exceptional Lie Algebras

We give here only the Cartan matrices of G_2 and F_4.

$$G_2 : \begin{vmatrix} 2 & -1 \\ -3 & 2 \end{vmatrix} \qquad (2.34)$$

$$F_4 : \begin{vmatrix} 2 & -1 & 0 & 0 \\ -1 & 2 & -2 & 0 \\ 0 & -1 & 2 & -1 \\ 0 & 0 & -1 & 2 \end{vmatrix}. \qquad (2.35)$$

Those of E_6, E_7 and E_8 can be found in Humphreys (1972) p. 59.

2.13   Real Forms of Complex Semisimple Lie Algebras

In applications, one is often interested in real forms of complex semi-simple Lie algebras. Commonly used real forms and their notation are given in Table 2.2.

Table 2.2 Real forms of complex semisimple Lie algebras

Cartan	Real forms	
A_l	$su(n)$	
	$su(p,q)$	$(p+q=n)$
	$sl(n,R)$	
	$su^*(2n)$	
B_l	$so(n)$	(n odd)
	$so(p,q)$	$(p+q=n)$
C_l	$sp(n)$	(n even)
	$sp(n,R)$	
	$sp(p,q)$	$(p+q=n)$
D_l	$so(n)$	(n even)
	$so(p,q)$	
	$so^*(n)$	

Table 2.3 Isomorphisms of complex semisimple Lie algebra

$A_1 \sim B_1 \sim C_1$
$B_2 \sim C_2$
$D_2 \sim A_1 \oplus A_1$
$A_3 \sim D_3.$

2.14 Isomorphisms of Complex Semisimple Lie Algebras

Isomorphisms of complex Lie algebras of low rank are given in Table 2.3.

2.15 Isomorphisms of Real Lie Algebras

Isomorphisms of real Lie algebras of low rank are given in Table 2.4.

2.16 Enveloping Algebra

Starting with a Lie algebra $g \ni X_\rho$, one can form the algebra composed of all products of elements

$$X_\rho$$

$$X_\rho X_\sigma$$

$$X_\rho X_\sigma X_\tau$$

$$\cdots \qquad (2.36)$$

Table 2.4 Isomorphisms of real Lie algebras

$A_1 \sim B_1 \sim C_1$	$su(2) \sim so(3) \sim sp(2) \sim su^{\star}(2)$
	$su(1,1) \sim so(2,1) \sim sp(2,R) \sim sl(2,R)$
$B_2 \sim C_2$	$so(5) \sim sp(4)$
	$so(4,1) \sim sp(2,2)$
	$so(3,2) \sim sp(4,R)$
$D_2 \sim A_1 \oplus A_1$	$so(4) \sim su(2) \oplus su(2) \sim so(3) \oplus so(3) \sim sp(2) \oplus sp(2)$
	$so^{\star}(4) \sim su(2) \oplus sl(2,R)$
	$so(3,1) \sim sl(2,C)$
	$so(2,2) \sim sl(2,R) \oplus sl(2,R)$
$A_3 \sim D_3$	$su(4) \sim so(6)$
	$su(3,1) \sim so^{\star}(6)$
	$su^{\star}(4) \sim so(5,1)$
	$sl(4,R) \sim so(3,3)$
	$su(2,2) \sim so(4,2)$

This algebra is called the enveloping algebra of g. The commutation relations of the X_ρ's among themselves define the Lie algebra g. The commutation relations of the enveloping algebra with the X_ρ's define a tensor algebra over g, $T(g)$.

2.17 Realizations of Lie Algebras

Lie algebras can be realized in various ways. Three of them have been widely used. In these realizations, elements are written in double index notation, $E_{\alpha\beta}$.

(i) Differential realization
  This is in terms of differential operators acting on functions $f(x_1, .., x_n)$

$$E_{\alpha\beta} = x_\alpha \frac{\partial}{\partial x_\beta} \quad . \tag{2.37}$$

Commutation relations of the algebra can be obtained from the basic commutation relations

$$\left[\frac{\partial}{\partial x_\alpha}, x_\beta \right] = \delta_{\alpha\beta} \tag{2.38}$$

Differential realizations will be discussed in Chap. 11.

(ii) Matrix realization

This is in terms of $n \times n$ matrices acting on column vectors $\begin{pmatrix} \cdots \\ \cdots \\ \cdots \end{pmatrix}$

$$E_{\alpha\beta} = \begin{pmatrix} & \vdots & \\ & \vdots & \\ -\,-\,-\,-\,- & 1 & -\,- \\ & \vdots & \\ & \vdots & \end{pmatrix} \qquad (2.39)$$

with unit entry on the β-th column and the α-th row. Commutation relations of the algebra are obtained from the basic commutation relations of matrices. Matrix realizations will be discussed in Chap. 12.

(iii) Boson   creation-annihilation   operator   realization   (often   called Jordan-Schwinger realization)

This is in terms of bilinear products of n boson creation, $b_\alpha^\dagger$, and annihilation, b_α, operators acting on a vacuum $|0\rangle$,

$$E_{\alpha\beta} = b_\alpha^\dagger b_\beta. \qquad (2.40)$$

The commutation relations of the algebra can be obtained from those of the creation and annihilation operators

$$\left[b_\alpha, b_\beta^\dagger\right] = \delta_{\alpha\beta} \quad , \qquad (2.41)$$

called Bose commutation relations. Realizations in terms of boson creation and annihilation operators will be discussed in Chap. 9.

For all three realizations, the commutation relations of the elements of the algebra $E_{\alpha\beta}$ are

$$[E_{\alpha\beta} \, , \, E_{\gamma\delta}] = \delta_{\beta\gamma} E_{\alpha\delta} - \delta_{\alpha\delta} E_{\gamma\beta} \qquad (2.42)$$

with $\alpha, \beta = 1, \ldots, n$. They define the Lie algebra of $u(n)$. Realizations of other algebras can be obtained by taking appropriate combinations of the elements $E_{\alpha\beta}$, since any Lie algebra is, by Ado's theorem, a subalgebra of $u(n)$.

2.18   Other Realizations of Lie Algebras

In addition to the three realizations of the previous section, others are possible and have been used for applications in physics. Two of these, particularly important in the description of fermionic systems, are:

(iv) Grassmann differential realization

This realization is in terms of Grassmann variables, θ_i ($i = 1, \ldots, n$), and their derivatives, $\frac{\partial}{\partial\theta_i}$. Grassmann variables are anticommuting variables satisfying

$$\theta_i \theta_j + \theta_j \theta_i = 0, \quad i, j = 1, \ldots, n. \tag{2.43}$$

The elements of the Lie algebra are the bilinear products

$$E_{ij} = \theta_i \frac{\partial}{\partial \theta_j}. \tag{2.44}$$

The commutation relations can be obtained from the basic commutation relations

$$\left[\frac{\partial}{\partial \theta_i}, \theta_j \right]_+ \equiv \frac{\partial}{\partial \theta_i} \theta_j + \theta_j \frac{\partial}{\partial \theta_i} = \delta_{ij}. \tag{2.45}$$

These realizations were first introduced by Martin in 1959 [I.L. Martin, Proc. Roy. Soc. A251, 536 (1959)] and later developed by several authors. They will not be discussed in these notes. A detailed account is given by Berezin (1987).

(v) Fermion creation-annihilation operators realization

This is in terms of bilinear products of n fermion creation, $a_i^\dagger$, and annihilation, a_i, operators acting on a vacuum $| 0 \rangle$,

$$E_{ij} = a_i^\dagger a_j, \tag{2.46}$$

where the creation and annihilation operators, $a_i^\dagger, a_i$, satisfy

$$[a_i, a_j^\dagger]_+ \equiv a_i a_j^\dagger + a_j^\dagger a_i = \delta_{ij}, \tag{2.47}$$

called Fermi commutation relations. These realizations will be discussed in Chap. 10.

For both realizations (iv) and (v), the commutation relations of the elements of the algebra E_{ij} are

$$\left[E_{ij}, E_{km} \right] = \delta_{jk} E_{im} - \delta_{im} E_{kj}, \tag{2.48}$$

with $i, j = 1, \ldots n$. The commutation relations (2.48) are identical to those given by (2.42). They define again the Lie algebra $u(n)$.

Chapter 3
Lie Groups

3.1 Groups of Transformations

A set of elements $A, B, C, \ldots$, forms a group G if it satisfies the following axioms:

Axiom 1. *Among the elements there is an element I such that*

$$AI = IA = A \tag{3.1}$$

This property is called identity.

Axiom 2. *The product AB gives another element C in the set*

$$AB = C \tag{3.2}$$

This property is called closure.

Axiom 3. *There exists an element A^{-1} such that*

$$A^{-1}A = AA^{-1} = I \tag{3.3}$$

This property is called inverse.

Axiom 4. *The order of multiplication is immaterial*

$$A(BC) = (AB)C \tag{3.4}$$

This property is called associativity.

Groups of transformations can be divided into discrete (finite and infinite) and continuous (finite and infinite). For discrete groups the number of elements is called the order of the group. For continuous groups, the number of parameters, to be

© Springer-Verlag Berlin Heidelberg 2015
F. Iachello, *Lie Algebras and Applications*, Lecture Notes in Physics 891,
DOI 10.1007/978-3-662-44494-8_3

described in the following sections, is called the order of the group. Both discrete and continuous groups are of importance in physics. Here we briefly describe some continuous groups (Lie groups) and their association with Lie algebras. A description of discrete groups is given by Hamermesh (1962).

The definition of groups of transformations given above is that used by physicists. For a mathematical definition of Lie groups, see Varadarajan (1984).

3.2 Groups of Matrices

Among the groups of transformations, particularly important are groups of square matrices

$$ A = \begin{pmatrix} \cdots \\ \cdots \\ \cdots \end{pmatrix} \qquad (n \times n). \tag{3.5} $$

These matrices satisfy all the axioms of a group:

1. The identity I is the unit matrix

$$ I = \begin{pmatrix} 1 & & & & 0 \\ & 1 & & & \\ & & \cdot & & \\ & & & \cdot & \\ & & & & \cdot \\ 0 & & & & 1 \end{pmatrix} \tag{3.6} $$

2. Matrix multiplication gives closure.
3. If det $\mid A \mid \neq 0$ an inverse A^{-1} exists.
4. Matrix multiplication gives associativity.

Groups of matrices can be written in terms of all number fields, R, C, Q, O. In these notes, we shall consider only groups of real and complex matrices. The matrix elements of the matrix A will be denoted by A_{ik}, with $i = $ row index and $k = $ column index. We shall also introduce real and complex vectors in n dimensions. The components of vectors will be denoted by x_i and z_i. Standard matrix notation will be used in this chapter (no covariant or contravariant indices).

3.3 Properties of Matrices

We begin by recalling in Table 3.1 some basic properties of matrices.

In this table, A^t denotes the transpose matrix, A^* the complex conjugate matrix, and $A^\dagger$ the hermitian conjugate matrix, $A^\dagger = (A^t)^*$.

Table 3.1 Matrix properties

$A = A^t$	Symmetric
$A = -A^t$	Skew symmetric
$A^t A = I$	Orthogonal
$A = A^*$	Real
$A = -A^*$	Imaginary
$A = A^\dagger$	Hermitian
$A = -A^\dagger$	Skew hermitian
$A^\dagger A = I$	Unitary

A group of transformations transforms the real or complex vector $\mathbf{x} \equiv (x_1, x_2, \ldots, x_n)$ or $\mathbf{z} \equiv (z_1, z_2, \ldots, z_n)$ into the real or complex vector $\mathbf{x}'$ or $\mathbf{z}'$. We shall consider both real and complex transformations

$$\mathbf{x}' = A\mathbf{x} \; ; \; x_i' = \sum_k A_{ik} x_k$$

$$\mathbf{z}' = B\mathbf{z} \; ; \; z_i' = \sum_k B_{ik} z_k \qquad (3.7)$$

where A_{ik} and B_{ik} are the matrix elements of the real and complex $n \times n$ matrices A and B.

3.4   Continuous Matrix Groups

1. General linear groups

   These are the most general linear transformations. They are denoted by

$$GL(n, C) \qquad r = 2n^2$$
$$GL(n, R) \qquad r = n^2. \qquad (3.8)$$

The number of real parameters that characterize the transformation is given next to its name. The number field R, C is also explicitly shown.

2. Special linear groups

   If, on the general linear transformation, the condition

$$\det | A | = +1 \qquad (3.9)$$

is imposed, the group is called special linear group, denoted by

$$SL(n, C) \qquad r = 2(n^2 - 1)$$
$$SL(n, R) \qquad r = n^2 - 1. \qquad (3.10)$$

3. Unitary groups

   Imposing the condition

$$A^\dagger A = I \tag{3.11}$$

one obtains the unitary groups

$$U(n, C) \equiv U(n) \qquad r = n^2$$
$$U(p, q; C) \equiv U(p, q) \qquad r = n^2. \tag{3.12}$$

They leave invariant the quantities

$$U(n) : \quad \sum_{i=1}^{n} z_i z_i^*$$

$$U(p, q) : \quad -\sum_{i=1}^{p} z_i z_i^* + \sum_{j=p+1}^{p+q} z_j z_j^* \tag{3.13}$$

Unitary groups are over complex numbers C. It has become common practice to delete the number field from the group notation, that is to use $U(n)$ instead of $U(n, C)$.

4. Special unitary groups

   The combination of the special condition with the unitary condition

$$A^\dagger A = I \quad , \qquad \det | A | = +1 \tag{3.14}$$

gives the special unitary groups

$$SU(n, C) \equiv SU(n) \qquad r = n^2 - 1$$
$$SU(p, q; C) \equiv SU(p, q) \quad r = n^2 - 1. \tag{3.15}$$

Again, the number field C is often deleted. For special unitary groups, there is an (anomalous) case, denoted by $SU^*(2n)$,

$$SU^*(2n) \quad r = (2n)^2 - 1 \tag{3.16}$$

defined by matrices

$$A = \begin{pmatrix} A_1 & A_2 \\ -A_2^* & A_1^* \end{pmatrix}$$

$A_1, A_2 = n \times n$ complex matrices with $Tr\, A_1 + Tr\, A_1^* = 0$. $\tag{3.17}$

5. Orthogonal groups

These groups are defined by the orthogonality condition

$$A^t A = I. \tag{3.18}$$

In applications in physics, they are usually over the real number field. The number field is often deleted in the notation and $O(n, R)$ is often denoted by $O(n)$

$$O(n, C) \quad r = n(n - 1)$$

$$O(n, R) \equiv O(n) \quad r = \frac{1}{2} n(n - 1). \tag{3.19}$$

They leave invariant the quantities

$$O(n, C) : \quad \sum_{i=1}^{n} z_i^2$$

$$O(n, R) \equiv O(n) : \quad \sum_{i=1}^{n} x_i^2 \tag{3.20}$$

In addition, one has the groups

$$O(p, q; C) \quad r = n(n - 1)$$

$$O(p, q; R) \quad r = \frac{1}{2} n(n - 1) \tag{3.21}$$

which leave invariant the quantities

$$O(p, q; C) : \quad -\sum_{i=1}^{p} z_i^2 + \sum_{j=p+1}^{p+q} z_j^2$$

$$O(p, q; R) \equiv O(p, q) : \quad -\sum_{i=1}^{p} x_i^2 + \sum_{j=p+1}^{p+q} x_j^2 \tag{3.22}$$

6. Special orthogonal groups

The combination of the special with the orthogonal condition

$$A^t A = I \quad \det | A | = +1 \tag{3.23}$$

gives the special orthogonal groups

$$SO(n, C) \quad r = n(n-1)$$

$$SO(n, R) \quad r = \frac{1}{2}n(n-1)$$

$$SO(p, q; C) \quad r = n(n-1)$$

$$SO(p, q; R) \quad r = \frac{1}{2}n(n-1). \tag{3.24}$$

Also here there is an (anomalous) case, called $SO^*(2n)$, described by matrices

$$A = \begin{pmatrix} A_1 & A_2 \\ -A_2^* & A_1^* \end{pmatrix}$$

$A_1, A_2 = n \times n$ complex matrices with $A_1 = -A_1^t$ and $A_2 = A_2^\dagger$. $\quad$ (3.25)

Real orthogonal groups are used both in quantum and in classical mechanics.

7. Symplectic groups

To define these groups, the vectors $\mathbf{x}$ and $\mathbf{y}$ are divided into two pieces, $\mathbf{x} = (x_1, \ldots, x_n; x_1', \ldots, x_n'), \mathbf{y} = (y_1, \ldots y_n; y_1', \ldots, y_n')$. Symplectic groups

$$Sp(2n, C) \quad r = 2n(2n+1)$$

$$Sp(2n, R) \quad r = \frac{1}{2}2n(2n+1) \tag{3.26}$$

are defined as those groups that leave invariant the quantity

$$\sum_{i=1}^{n} \left(x_i y_i' - y_i x_i' \right) \tag{3.27}$$

where the vectors can be either real or complex. If the unitary condition is imposed

$$A^\dagger A = I \tag{3.28}$$

the group is called unitary symplectic

$$USp(2n, C) \equiv Sp(2n) \qquad r = \frac{1}{2} 2n(2n+1) \tag{3.29}$$

and is often denoted by $Sp(2n)$.

Both real and complex symplectic groups are used in quantum mechanics, while real symplectic groups are used in classical mechanics (canonical transformations and Hamilton's equations).

3.5 Examples of Groups of Transformations

3.5.1 The Rotation Group in Two Dimensions, SO(2)

As a first example we consider the rotation group in two dimensions $SO(2) \equiv SO(2, R)$. Under a general linear real transformation the two coordinates x, y (used here to conform with usual physics notation) transform as

$$x' = a_{11} \, x \, + \, a_{12} \, y$$
$$y' = a_{21} \, x \, + \, a_{22} \, y \tag{3.30}$$

The corresponding group, $GL(2, R)$, is a four parameter group. The invariance of $x^2 + y^2$

$$a_{11}^2 \, x^2 + a_{12}^2 \, y^2 + 2 \, a_{11} a_{12} \, xy + a_{21}^2 x^2 + a_{21}^2 y^2 + 2 \, a_{21} a_{22} \, xy = x^2 + y^2 \tag{3.31}$$

gives three conditions

$$a_{11}^2 + a_{21}^2 = 1$$
$$2a_{11} a_{12} + 2a_{21} a_{22} = 0$$
$$a_{22}^2 + a_{12}^2 = 1. \tag{3.32}$$

This leaves only one parameter.

Example 1. The group SO(2) is a one parameter group

The parameter can be chosen as the angle of rotation, φ,

$$x' = (\cos \varphi) \, x - (\sin \varphi) \, y$$
$$y' = (\sin \varphi) \, x + (\cos \varphi) \, y \tag{3.33}$$

as shown in Fig. 3.1.

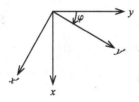

Fig. 3.1 The angle φ that parametrizes $SO(2)$

3.5.2 The Lorentz Group in One Plus One Dimension, $SO(1, 1)$

A group closely related to the rotation group is the Lorentz group $SO(1, 1) \equiv SO(1, 1; R)$. The general linear real transformation in space-time, x, t can be written

$$x' = a_{11}x + a_{12}t$$
$$t' = a_{21}x + a_{22}t. \tag{3.34}$$

Imposing the condition $x^2 - t^2 =$ invariant, leaves a one parameter group.

Example 2. The group SO(1,1) is a one parameter group

A convenient parametrization is in term of the boost, ϑ.

$$x' = (\cosh \vartheta)x + (\sinh \vartheta)t$$
$$t' = (\sinh \vartheta)x + (\cosh \vartheta)t. \tag{3.35}$$

By comparing with the previous subsection, one can see that the invariant forms are different (Fig. 3.2).

The group $SO(2)$ which leaves invariant the form $x^2 + y^2$ is said to be compact, while the group $SO(1, 1)$ which leaves invariant $x^2 - t^2$ is said to be non-compact. The notation is such that the number of plus or minus signs is indicated in $SO(p, q)$.

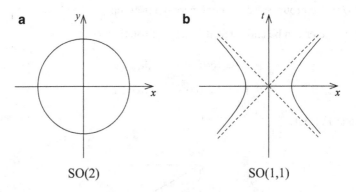

Fig. 3.2 Invariant forms of **a** $SO(2)$ and **b** $SO(1, 1)$

3.5.3 The Rotation Group in Three Dimensions, $SO(3)$

As another example consider the rotation group in three dimensions $SO(3) \equiv SO(3, R)$. Under a general linear transformation, $GL(3, R)$, the coordinates x, y, z transform as

$$
\begin{aligned}
x' &= a_{11} x + a_{12} y + a_{13} z \\
y' &= a_{21} x + a_{22} y + a_{23} z \\
z' &= a_{31} x + a_{32} y + a_{33} z
\end{aligned} \tag{3.36}
$$

This is a nine parameter group. Orthogonality

$$
x'^2 + y'^2 + z'^2 = x^2 + y^2 + z^2 \tag{3.37}
$$

gives six conditions. We thus have a three parameter group.

Example 3. The group $SO(3)$ is a three parameter group

A convenient parametrization is in terms of Euler angles, φ, ϑ, ψ,

$$
\begin{pmatrix}
\cos\varphi \cos\vartheta \cos\psi - \sin\varphi \sin\psi & -\cos\varphi \cos\vartheta \sin\psi - \sin\varphi \cos\psi & \cos\varphi \sin\vartheta \\
\sin\varphi \cos\vartheta \cos\psi + \cos\varphi \sin\psi & -\sin\varphi \cos\vartheta \sin\psi + \cos\varphi \cos\psi & -\sin\varphi \sin\vartheta \\
-\sin\vartheta \cos\psi & \sin\vartheta \sin\psi & \cos\vartheta
\end{pmatrix} \tag{3.38}
$$

as shown in Fig. 3.3. The rotation matrices (3.38) are usually denoted by $R(\varphi, \vartheta, \psi)$.

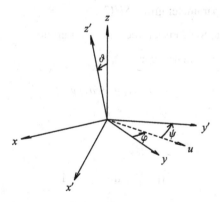

Fig. 3.3 The Euler angles φ, ϑ, ψ that parametrize $SO(3)$

3.5.4 The Special Unitary Group in Two Dimensions, $SU(2)$

This group is denoted by $SU(2) \equiv SU(2,C)$. Under a general linear complex transformation, $GL(2,C)$, the complex quantities, u, v, called a spinor, transform as

$$u' = a_{11} u + a_{12} v$$
$$v' = a_{21} u + a_{22} v. \tag{3.39}$$

This is a eight parameter group. Call the matrix of the transformation A

$$A = \begin{pmatrix} a_{11} & a_{12} \\ a_{21} & a_{22} \end{pmatrix} \qquad A^{\dagger} = \begin{pmatrix} a_{11}^{*} & a_{21}^{*} \\ a_{12}^{*} & a_{22}^{*} \end{pmatrix} \tag{3.40}$$

Unitarity, $A^{\dagger} A = 1$, gives four conditions

$$a_{11}^{*} a_{11} + a_{21}^{*} a_{21} = 1$$
$$a_{11}^{*} a_{12} + a_{21}^{*} a_{22} = 0$$
$$a_{12}^{*} a_{11} + a_{22}^{*} a_{21} = 0$$
$$a_{12}^{*} a_{12} + a_{22}^{*} a_{22} = 1 \tag{3.41}$$

The corresponding group, $U(2)$, is a four parameter group. If one imposes a further condition $\det \mid A \mid = +1$, that is

$$a_{11} a_{22} - a_{12} a_{21} = 1 \tag{3.42}$$

one obtains the three parameter group $SU(2)$.

Example 4. The group SU(2) is a three parameter group

This group can be parametrized as

$$u' = a_{11} u + a_{12} v$$
$$v' = -a_{12}^{*} u + a_{11}^{*} v \tag{3.43}$$

with

$$a_{11} a_{11}^{*} + a_{12} a_{12}^{*} = 1. \tag{3.44}$$

3.5.5 Relation Between SO(3) and SU(2)

Both $SO(3)$ and $SU(2)$ are three parameter groups. It is of importance to find their relationship. Consider the following combination of the complex spinor u, v

$$x_1 = u^2 ; \quad x_2 = uv ; \quad x_3 = v^2. \tag{3.45}$$

These combinations transform as

$$
\begin{aligned}
x_1' &= u'^2 = a_{11}^2 x_1 + 2a_{11} a_{12} x_2 + a_{12}^2 x_3 \\
x_2' &= u'v' = -a_{11} a_{12}^* x_1 + (a_{11} a_{11}^* - a_{12} a_{12}^*) x_2 + a_{11}^* a_{12} x_3 \\
x_3' &= v'^2 = a_{12}^{*2} x_1 - 2a_{11}^* a_{12}^* x_2 + a_{11}^{*2} x_3
\end{aligned} \tag{3.46}
$$

By introducing the coordinates x, y, z

$$x = (x_1 - x_3)/2 ; \quad y = (x_1 + x_3)/2i ; \quad z = x_2 \tag{3.47}$$

one can see that they transform as

$$
\begin{aligned}
x' &= \tfrac{1}{2}(a_{11}^2 - a_{12}^{*2} - a_{12}^2 + a_{11}^{*2})x + \tfrac{i}{2}(a_{11}^2 - a_{12}^{*2} + a_{12}^2 - a_{11}^{*2})y \\
&\quad + (a_{11}a_{12} + a_{11}^* a_{12}^*)z \\
y' &= \tfrac{-i}{2}(a_{11}^2 - a_{12}^{*2} - a_{12}^2 + a_{11}^{*2})x + \tfrac{1}{2}(a_{11}^2 - a_{12}^{*2} + a_{12}^2 + a_{11}^{*2})y \\
&\quad - i\,(a_{11}a_{12} - a_{11}^* a_{12}^*)z \\
z' &= -(a_{11}^* a_{12} + a_{11} a_{12}^*)x + i\,(a_{11}^* a_{12} - a_{11}a_{12}^*)y \\
&\quad + (a_{11} a_{11}^* - a_{12} a_{12}^*)z
\end{aligned} \tag{3.48}
$$

This is a real orthogonal transformation in three dimensions, satisfying

$$x'^2 + y'^2 + z'^2 = x^2 + y^2 + z^2. \tag{3.49}$$

Thus $SU(2)$ and $SO(3)$ are related by a change of variables. In order to elucidate the correspondence between $SU(2)$ and $SO(3)$, we consider a rotation of an angle α around the z-axis. By inserting the values $a_{11} = e^{i\alpha/2}, a_{12} = 0$ in the appropriate formulas, we see that this rotation is characterized by matrices

$$
\begin{array}{cc}
SU(2) & SO(3)
\end{array}
$$

$$
\begin{pmatrix} e^{i\alpha/2} & 0 \\ 0 & e^{-i\alpha/2} \end{pmatrix}
\qquad
\begin{pmatrix} \cos\alpha & -\sin\alpha & 0 \\ \sin\alpha & \cos\alpha & 0 \\ 0 & 0 & 1 \end{pmatrix}
\tag{3.50}
$$

A generic rotation, by angles α, β, γ is instead characterized by matrices

$$SU(2) \qquad\qquad\qquad SO(3)$$

$$\begin{pmatrix} \cos\frac{\beta}{2}e^{\frac{i}{2}(\alpha+\gamma)} & \sin\frac{\beta}{2}e^{-\frac{i}{2}(\alpha-\gamma)} \\ -\sin\frac{\beta}{2}e^{\frac{i}{2}(\alpha-\gamma)} & \cos\frac{\beta}{2}e^{-\frac{i}{2}(\alpha+\gamma)} \end{pmatrix} \quad R(\alpha,\beta,\gamma) \qquad (3.51)$$

where $R(\alpha,\beta,\gamma)$ is given in (3.38). For no rotation, $R(0,0,0)$, the correspondence is

$$SU(2) \quad SO(3)$$

$$\begin{pmatrix} 1 & 0 \\ 0 & 1 \end{pmatrix} \quad \begin{pmatrix} 1 & 0 & 0 \\ 0 & 1 & 0 \\ 0 & 0 & 1 \end{pmatrix} \qquad (3.52)$$

while for rotation of 2π, $R(2\pi,0,0)$, the correspondence is

$$SU(2) \qquad SO(3)$$

$$\begin{pmatrix} -1 & 0 \\ 0 & -1 \end{pmatrix} \quad \begin{pmatrix} 1 & 0 & 0 \\ 0 & 1 & 0 \\ 0 & 0 & 1 \end{pmatrix} \qquad (3.53)$$

One can see that there is a two-to-one correspondence, called a homomorphic mapping of $SU(2)$ into $SO(3)$, denoted by $SU(2) \approx SO(3)$. One says that $SU(2)$ is the universal covering group of $SO(3)$.

3.6   Other Important Groups of Transformations

An important class of transformations is formed by the combination of the translation group with the general linear group and its subgroups. These groups are still Lie groups but the associated Lie algebras are non-semisimple.

3.6.1   Translation Group, $T(n)$

Translations in n-dimensions form a group. Under a translation $\mathbf{a}$, the new coordinates are

$$\mathbf{x}' = \mathbf{x} + \mathbf{a}; \quad x_i' = x_i + a_i \quad (i = 1, \ldots, n). \qquad (3.54)$$

The translation group is a n parameter group.

3.6.2 Affine Group, $A(n)$

General linear transformations with $\det |A| \neq 0$ plus translations form a group, called the affine group, $A(n)$, with

$$\mathbf{x}' = A\mathbf{x} + \mathbf{a}; \quad x'_i = \sum_k A_{ik} x_k + a_i \quad (i = 1, \ldots, n). \tag{3.55}$$

This group is the semidirect product of the general linear group and the translation group

$$A(n) = T(n) \otimes_s GL(n). \tag{3.56}$$

The number of parameters of $A(n)$ for real transformations is $n^2 + n$.

Matrix representations of the affine group can be constructed in terms of $(n + 1) \times (n + 1)$ matrices

$$\begin{pmatrix} A & \mathbf{a} \\ 0 & 1 \end{pmatrix}. \tag{3.57}$$

3.6.3 Euclidean Group, $E(n)$

Rotations plus translations in an n-dimensional space form a group, called the Euclidean group, $E(n)$. A vector $\mathbf{x}$ transforms under $E(n)$ as

$$\mathbf{x}' = R\mathbf{x} + \mathbf{a}; \quad x'_i = \sum_k R_{ik} x_k + a_i, \tag{3.58}$$

where R_{ik} is the rotation matrix and a_i are the components of the translation vector. The Euclidean group is the semi-direct product of $SO(n)$ and $T(n)$

$$E(n) = T(n) \otimes_s SO(n). \tag{3.59}$$

A case of particular interest is

$$E(3) = T(3) \otimes_s SO(3). \tag{3.60}$$

The Lie algebra $e(n)$ associated with $E(n)$ are the semidirect sums

$$e(n) = t(n) \oplus_s so(n). \tag{3.61}$$

The algebra $e(2)$ is given as an example in (1.37).

Since $E(n)$ is a subgroup of $A(n)$, matrix representations of $E(n)$ can be constructed as in (3.57). The number of parameters of $E(n)$ is $\frac{n(n-1)}{2} + n$.

3.6.4   Poincare' Group, $P(n)$

Lorentz transformations plus translations form a group, called the Poincare' group, $P(n)$. A vector $\mathbf{x}'$ transforms under $P(n)$ as

$$\mathbf{x}' = L\mathbf{x} + \mathbf{a}; \quad x_\mu = \sum_\nu L_\mu^\nu x_\nu + a_\mu, \tag{3.62}$$

where L_μ^ν are Lorentz transformations and a_μ are the components of the translation. This group is the semidirect product of $SO(p, q)$ and $T(p, q)$, with $p + q = n$,

$$P(n) = T(p, q) \otimes_s SO(p, q), \quad p + q = n. \tag{3.63}$$

A case of particular interest is

$$P(4) = T(3, 1) \otimes_s SO(3, 1). \tag{3.64}$$

This group is also denoted by $ISO(3, 1) \equiv P(4)$ or the inhomogeneous Lorentz group.

From Poincare' transformations, by the process of contraction discussed in Sect. 1.17, one can obtain Galilean transformations

$$\mathbf{x}' = R\mathbf{x} + \mathbf{v}t + \mathbf{a}; \quad x_i' = \sum_k R_{ik} x_k + v_i t + a_i, \tag{3.65}$$

where v_i are the components of the velocity vector $\mathbf{v}$ and a_i the components of the translation vector $\mathbf{a}$.

Matrix representations of $P(n)$ can also be constructed as in (3.57).

Another important class of transformations is formed by the combination of dilatations and affine transformations.

3.6.5   Dilatation Group, $D(1)$

Scale transformations form a one parameter group, called the dilatation group,

$$D(1): \quad x'^\mu = \rho x^\mu. \tag{3.66}$$

3.6.6 Special Conformal Group, $C(n)$

The set of *non-linear* transformations

$$x'^{\mu} = \left(x^{\mu} + c^{\mu}x^2\right)/\sigma(x)$$

$$\sigma(x) = 1 + 2c^{\nu}x_{\nu} + c^2x^2, \tag{3.67}$$

form a group, called the special conformal group, $C(n)$. In four dimensions, the group $C(4)$ has four parameters, $c_{\mu}(\mu = 0, 1, 2, 3)$.

3.6.7 General Conformal Group, $GC(n)$

The set of Lorentz transformations plus translations plus dilatations plus special conformal transformations form a group, the General Conformal Group, $GC(n)$, or simply the Conformal Group. In four dimensions, the number of parameters of $GC(4)$ is: 10 for the Poincare' group $ISO(3, 1) \equiv P(4)$, 1 for the dilatation, $D(1)$, and 4 for the special conformal transformations, $C(4)$, for a total of 15.

The group $GC(4)$ is isomorphic to $SO(4, 2)$. It is possible to introduce a six-dimensional space and realize the conformal group linearly in this space. A differential realization of the elements of the Lie algebra $so(4, 2)$ associated with the Lie group $SO(4, 2)$ is

$$\begin{array}{ll} M_{\mu\nu} = x_{\mu}\partial_{\nu} - x_{\nu}\partial_{\mu} & SO(3, 1) \\ P_{\mu} = \partial_{\mu} & T(3, 1) \\ K_{\mu} = 2x_{\mu}x^{\nu}\partial_{\nu} - x^2\partial_{\mu} & C(4) \\ D = x^{\nu}\partial_{\nu} & D(1), \end{array} \tag{3.68}$$

with $\mu, \nu = 0, 1, 2, 3$. Conformal transformations can be written as linear transformations in a six-dimensional space with coordinates $\eta^{\mu} = kx^{\mu}, k, \lambda = kx^2$. Dilatations and special conformal transformations acting in this space are

$$D(1): \quad \eta'^{\mu} = \eta^{\mu}, \quad k' = \rho^{-1}k, \quad \lambda' = \rho\lambda$$

$$C(4): \quad \eta'^{\mu} = \eta^{\mu} + c^{\mu}\lambda, \quad k' = -2c_{\nu}\eta^{\nu} + k + c^2\lambda, \quad \lambda' = \lambda, \tag{3.69}$$

while Poincare' transformations act as in Sect. 3.6.4.

Chapter 4
Lie Algebras and Lie Groups

4.1 The Exponential Map

The relationship between Lie algebras and Lie groups is of great importance. Let the Lie algebra be g and the corresponding Lie group G. The relation is

$$\text{Lie algebra} \qquad g \ni X_i \quad (i = 1,\ldots,r) \qquad (4.1)$$

$$\text{Lie group} \qquad G \ni \exp\left(\sum_{i=1}^{r} \alpha_i X_i\right) \qquad (4.2)$$

where the α_i's are the parameters of the group and the sum goes over the order of the group. (The α_i's here should not be confused with the α_i's in Chap. 2 where they denote the components of a root vector). This relationship is called an exponential map and denoted by

$$
\begin{array}{c}
g \\
\downarrow \exp \\
G
\end{array}
\qquad (4.3)
$$

Example 1. The Lie group SO(3) is

$$A(\alpha_1, \alpha_2, \alpha_3) = e^{\alpha_1 X_1 + \alpha_2 X_2 + \alpha_3 X_3} \qquad (4.4)$$

4.2 Definition of Exp

The exponentiation is defined through a power series expansion. For rank one algebras, with only one element X and one parameter α

© Springer-Verlag Berlin Heidelberg 2015
F. Iachello, *Lie Algebras and Applications*, Lecture Notes in Physics 891,
DOI 10.1007/978-3-662-44494-8_4

$$e^{\alpha X} = 1 + \alpha X + \frac{\alpha^2 X^2}{2!} + \cdots = \sum_{p=0}^{\infty} \frac{(\alpha X)^p}{p!} \qquad (4.5)$$

The infinitesimal group element is obtained by keeping only the linear term in the expansion

$$e^{\alpha x} \underset{\alpha \to 0}{\longrightarrow} 1 + \alpha X \qquad (4.6)$$

For algebras of larger rank, one needs to exponentiate non-commuting elements. It is convenient to use matrices.

4.3 Matrix Exponentials

Let A be a $n \times n$ matrix. Then

$$e^A = I + A + \frac{A^2}{2!} + \cdots \qquad (4.7)$$

Some properties of matrix exponentials are:

1. The exponential e^A converges if the matrix elements $| a_{ij} |$ have an upper bound, that is the group is compact.
2. If A and B commute, then

$$e^{A+B} = e^A e^B. \qquad (4.8)$$

3. If B can be inverted, then

$$B e^A B^{-1} = e^{BAB^{-1}} \qquad (4.9)$$

4. If $\lambda_1, \lambda_2, \ldots, \lambda_n$ are eigenvalues of A, then

$$e^{\lambda_1}, \ldots, e^{\lambda_n} \qquad (4.10)$$

are eigenvalues of e^A.
5. The exponential series satisfies

$$e^{A^*} = (e^A)^* \quad (e^{A^t}) = (e^A)^t$$
$$e^{A^\dagger} = (e^A)^\dagger \quad e^{-A} = (e^A)^{-1} \qquad (4.11)$$

6. The determinant of e^A is e^{trA}.
7. If A is skew symmetric, e^A is orthogonal. If A is skew hermitian, e^A is unitary.

8. The following formula (Campbell–Hausdorff) applies

$$e^{-A} B e^{A} = B + \frac{1}{1!}[B, A] + \frac{1}{2!}[[B, A], A] + \dots \tag{4.12}$$

4.4 More on Exponential Maps

The exponential map produces a particular parametrization of the group, that connected with the identity element.

Example 2. Lie group SO(3)

Denote by α_1 the angle of rotation about x, α_2 about y and α_3 about z. The rotation matrix $A(\alpha_1, \alpha_2, \alpha_3)$ in terms of these angles is

$$
\begin{pmatrix}
\cos\alpha_2 \cos\alpha_3 & -\sin\alpha_1 \sin\alpha_2 \cos\alpha_3 + \cos\alpha_1 \sin\alpha_3 \\
-\cos\alpha_2 \sin\alpha_3 & \cos\alpha_1 \cos\alpha_3 + \sin\alpha_1 \sin\alpha_2 \sin\alpha_3 \\
-\sin\alpha_2 & -\sin\alpha_1 \cos\alpha_2 \\
& \\
& \cos\alpha_1 \sin\alpha_2 \cos\alpha_3 + \sin\alpha_1 \sin\alpha_3 \\
& -\cos\alpha_1 \sin\alpha_2 \sin\alpha_3 + \sin\alpha_1 \cos\alpha_2 \\
& \cos\alpha_1 \cos\alpha_2
\end{pmatrix},
\tag{4.13}
$$

where $-\pi \le \alpha_1 \le \pi, -\pi \le \alpha_2 \le \pi, -\frac{\pi}{2} \le \alpha_3 \le \frac{\pi}{2}$. This matrix can be obtained from the exponential map (4.4) with

$$X_{\alpha_1} = \begin{pmatrix} 0 & 0 & 0 \\ 0 & 0 & -1 \\ 0 & 1 & 0 \end{pmatrix}, \quad X_{\alpha_2} = \begin{pmatrix} 0 & 0 & 1 \\ 0 & 0 & 0 \\ -1 & 0 & 0 \end{pmatrix}, \quad X_{\alpha_3} = \begin{pmatrix} 0 & -1 & 0 \\ 1 & 0 & 0 \\ 0 & 0 & 0 \end{pmatrix}, \tag{4.14}$$

satisfying the commutation relations of the Lie algebra $so(3)$ (1.9)

$$\left[X_{\alpha_i}, X_{\alpha_j} \right] = \varepsilon_{ijk} X_{\alpha_k} \tag{4.15}$$

where ε_{ijk} is the antisymmetric rank-3 tensor. The elements of the Lie algebra $X_{\alpha_i} (i = 1, 2, 3)$ are called generators of the group.

Physicists often use other parametrizations. For example, in the Euler angle parametrization (3.38), an operator

$$R(\varphi, \vartheta, \psi) = R_{z'}(\psi) R_u(\vartheta) R_z(\varphi) = e^{-i\psi J_{z'}} e^{-i\vartheta J_u} e^{-i\varphi J_z}, \tag{4.16}$$

is introduced, where the axes z', u, z are shown in Fig. 3.3 (Messiah 1958). By a series of transformations R can be brought to the form

$$R\left(\varphi, \vartheta, \psi\right) = e^{-i\varphi J_z} e^{-i\vartheta J_y} e^{-i\psi J_z}, \tag{4.17}$$

where J_x, J_y, J_z satisfy the commutation relations (1.18)

$$\left[J_x, J_y\right] = iJ_z, \quad \left[J_y, J_z\right] = iJ_x, \quad \left[J_z, J_x\right] = iJ_y. \tag{4.18}$$

Although this expression is useful in practical calculations, it is not an exponential map, since it is not connected with the identity element, and therefore it is not a parametrization of the group.

Example 3. Lie group SU(2)

The matrix parametrization $A(\alpha_1, \alpha_2, \alpha_3)$ of $SU(2)$ in terms of the angles $\alpha_1, \alpha_2, \alpha_3$ is

$$\begin{pmatrix} (\cos\alpha_1 \cos\alpha_2 + i \sin\alpha_1 \sin\alpha_2)e^{i\alpha_3} & -\cos\alpha_1 \sin\alpha_2 + i \sin\alpha_1 \cos\alpha_2 \\ \cos\alpha_1 \sin\alpha_2 + i \sin\alpha_1 \cos\alpha_2 & (\cos\alpha_1 \cos\alpha_2 - i \sin\alpha_1 \sin\alpha_2)e^{-i\alpha_3} \end{pmatrix}, \tag{4.19}$$

where $-\pi \le \alpha_1 \le \pi, -\pi \le \alpha_2 \le \pi, 0 \le \alpha_3 \le \pi$. This matrix can be obtained from the exponential map (4.4) with

$$X_1 = \begin{pmatrix} 0 & i \\ i & 0 \end{pmatrix}, \quad X_2 = \begin{pmatrix} 0 & -1 \\ 1 & 0 \end{pmatrix}, \quad X_3 = \begin{pmatrix} i & 0 \\ 0 & -i \end{pmatrix}, \tag{4.20}$$

satisfying the commutation relations of $su(2) \sim so(3)$, (1.9) and (4.15).

4.5   Infinitesimal Transformations

Infinitesimal transformations can be simply obtained from the exponential map by expanding the exponential and keeping only the first order terms.

Example 4. Infinitesimal $SO(3)$ rotation around z

This infinitesimal rotation is obtained from $A(\alpha_1, \alpha_2, \alpha_3)$ of (4.13) by letting $\alpha_1 = 0, \alpha_2 = 0, \alpha_3 = \varepsilon$. One obtains as $\varepsilon \to 0$,

$$A(0, 0, \varepsilon) = \begin{pmatrix} 1 & \varepsilon & 0 \\ -\varepsilon & 1 & 0 \\ 0 & 0 & 1 \end{pmatrix}. \tag{4.21}$$

By acting with $A(0, 0, \varepsilon)$ on a vector with components x, y, z, one obtains

$$A(0, 0, \varepsilon) \begin{pmatrix} x \\ y \\ z \end{pmatrix} = \begin{pmatrix} x + \varepsilon y \\ -\varepsilon x + y \\ z \end{pmatrix}. \tag{4.22}$$

Chapter 5
Homogeneous and Symmetric Spaces (Coset Spaces)

5.1 Definitions

Consider an algebra g with elements $X_i (i = 1, \ldots, r)$, $g \ni X_i$, and its associated group G obtained from g by exponentiation (4.2), $G \ni \exp\left(\sum_i \alpha_i X_i\right)$, where α_i are the parameters of the group. Consider a topological space Γ, with points denoted by γ. The group G transforms any point γ of Γ into another point γ'. The group G is called a topological (left) transformation group on Γ if

Axiom 1. *With each $G_\alpha \in G$ there is associated a homeomorphism $\gamma \to G\gamma$ of Γ into Γ* (schematically shown in Fig. 5.1)

Axiom 2. *The identity element I of G is the identity homeomorphism of Γ*

Axiom 3. *The mapping $\gamma \to G\gamma$ of $G \times \Gamma$ into Γ is continuous*

Axiom 4. *$(G_1 G_2)\gamma = G_1(G_2\gamma)$ for $G_1, G_2 \in G$ and $\gamma \in \Gamma$*

It is said that G acts transitively on Γ if for every pair of points $\gamma_1, \gamma_2 \in \Gamma$ there exists an element G such that $\gamma_2 = G\gamma_1$, and that G acts effectively on Γ if I is the only element that leaves each $\gamma \in \Gamma$ fixed. The space Γ is called a homogeneous space.

A realization of homogeneous spaces is provided by the quotient spaces G/H. If G is a connected Lie group and H a compact subgroup of G, $G \supset H$, the quotient space G/H is called a globally symmetric Riemannian space (also called a coset space).

© Springer-Verlag Berlin Heidelberg 2015
F. Iachello, *Lie Algebras and Applications*, Lecture Notes in Physics 891,
DOI 10.1007/978-3-662-44494-8_5

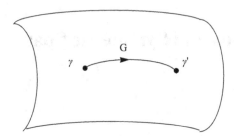

Fig. 5.1 Transformation of γ into γ' by G

Table 5.1 Irreducible globally symmetric Riemannian spaces whose transformation group is a simple connected Lie group

Compact	Non-compact	Rank	Dimension
$SU(n)/SO(n)$	$SL(n, R)/SO(n)$	$n-1$	$(n-1)(n+2)/2$
$SU(2n)/Sp(n)$	$SU^*(2n)/Sp(n)$	$n-1$	$(n-1)(2n+1)$
$SU(p+q)/S(U(p) \otimes U(q))$	$SU(p,q)/S(U(p) \otimes U(q))$	$\min(p,q)$	$2pq$
$SO(p+q)/SO(p) \otimes SO(q)$	$SO(p,q)/SO(p) \otimes SO(q)$	$\min(p,q)$	pq
$SO(2n)/U(n)$	$SO^*(2n)/U(n)$	$[n/2]$	$n(n-1)$
$Sp(2n)/U(n)$	$Sp(2n, R)/U(n)$	n	$n(n+1)$
$Sp(p+q)/Sp(p) \otimes Sp(q)$	$Sp(p,q)/Sp(p) \otimes Sp(q)$	$\min(p,q)$	$4pq$

5.2 Cartan Classification

Irreducible globally symmetric spaces were classified by Cartan (1926, 1927). They are given in Table 5.1. Cartan also classified symmetric spaces associated with the exceptional groups, not included in Table 5.1. This table gives only irreducible globally symmetric Riemannian spaces of Type I and Type III in Cartan's classification. In addition, there are also two other classes (Type II and IV) not reported here. Symmetric spaces G/H with non-compact stability groups have also been classified. A complete list is given in Barut and Rączka (1986).

5.3 How to Construct Coset Spaces

Given an algebra g and a subalgebra h of g, $g \supset h$, divide g into

$$g = h \oplus p. \qquad (5.1)$$

The algebra h is called the stability algebra, g/h the factor algebra, and p the remainder, not closed with respect to commutation. The number of elements of p gives the topological dimension of the space, last column in Table 5.1, and the dimension of the maximal abelian subalgebra of p gives the rank of the space.

The decomposition (5.1) has a counterpart in the Lie group, called a coset decomposition. Let G be the group associated with g, H the group associated with h, and $P = \exp p$,

$$G = \exp g$$
$$H = \exp h \tag{5.2}$$
$$P = \exp p.$$

Then

$$G = PH, \quad G = HP \tag{5.3}$$

are called the left and right coset decompositions respectively. The Reimannian space G/H is the parameter space of the coset.

Riemannian spaces especially important in physics are given in Table 5.2.

Example 1. The Riemannian space

$$U(6)/U(5) \otimes U(1) \tag{5.4}$$

has 10 variables (5 complex variables).

Example 2. The Riemannian space

$$SO(3)/SO(2) \tag{5.5}$$

has 2 real variables.

The spaces $U(n)/U(n-1) \otimes U(1)$ are useful when describing systems of bosons. The spaces $SO(n)/SO(n-1)$ are useful in quantum mechanics. An explicit construction of symmetric Riemannian spaces is given in the following Chap. 13.

Two other important Riemannian spaces related to those of Table 5.2 are $U(n,1)/U(n) \otimes U(1)$ and $SO(n,1)/SO(n)$. Their properties are listed in Table 5.3.

Table 5.2 Two important Riemannian spaces

Space	Rank	Dimension	Variables
$U(n)/U(n-1) \otimes U(1)$	1	$2(n-1)$	$(n-1)$-complex
$SO(n)/SO(n-1)$	1	$(n-1)$	$(n-1)$-real

Table 5.3 Two important Riemannian spaces of non-compact groups

Space	Rank	Dimension	Variables
$U(n,1)/U(n) \otimes U(1)$	1	$2n$	n-complex
$SO(n,1)/SO(n)$	1	n	n-real

Example 3. The Riemannian space

$$U(5, 1)/U(5) \otimes U(1) \tag{5.6}$$

has 10 variables (5 complex variables).

Example 4. The Riemannian space

$$SO(3, 1)/SO(3) \tag{5.7}$$

has 3 real variables.

The spaces $SO(n, 1)/SO(n)$ are particularly useful in relativistic quantum mechanics. An explicit construction of symmetric Riemannian spaces associated with the Lorentz groups $SO(n, 1)$ is given in Chap. 13.

Chapter 6
Irreducible Bases (Representations)

6.1 Definitions

An irreducible basis (*IRB*) is the basis for the representations of the algebra, g, (and the associated group, G), and the basis upon which the elements of the algebra, $X_\rho \in g$, act. It will be denoted by $\mathcal{B}$.

Also, if $\mathcal{V}$ is a linear vector space,

$$\mathcal{B} : \mathcal{V} \oplus (\mathcal{V} \oplus \mathcal{V}) \oplus (\mathcal{V} \oplus \mathcal{V} \oplus \mathcal{V}) \oplus \ldots \qquad (6.1)$$

The meaning of the term irreducible is that any element X_ρ acting on $\mathcal{B}$ does not lead out of $\mathcal{B}$.

6.2 Abstract Characterization

Irreducible representations are characterized by a set of labels (often called quantum numbers). For a semisimple Lie algebra g, the number of labels is the rank of the algebra, l, which is also the number of Cartan commuting elements. The number of labels and their notation in these lecture notes are shown in Table 6.1.

We shall consider here representations of the classical compact algebras, $su(n)$, $so(n)$, $sp(n)$. There are two types of representations: tensor representations and spinor representations.

6.3 Irreducible Tensors

6.3.1 Irreducible Tensors with Respect to GL(n)

In order to illustrate the notion of irreducible tensors we consider here two vectors $\mathbf{x} \equiv (x_1, \ldots, x_n)$ and $\mathbf{y} \equiv (y_1, \ldots, y_n)$. Under $GL(n)$ they transform as

© Springer-Verlag Berlin Heidelberg 2015
F. Iachello, *Lie Algebras and Applications*, Lecture Notes in Physics 891,
DOI 10.1007/978-3-662-44494-8_6

Table 6.1 Labels of irreducible representations

Name		Number of labels	Labels
$u(n)$		n	$[\lambda_1, \lambda_2, \ldots, \lambda_n]$
$su(n)$		$n-1$	$[\lambda_1, \lambda_2, \ldots, \lambda_{n-1}]$
$so(n)$	(n odd)	$v = (n-1)/2$	$[\mu_1, \mu_2, \ldots, \mu_v]$
$sp(n)$	(n even)	$v = n/2$	$[\mu_1, \mu_2, \ldots, \mu_v]$
$so(n)$	(n even)	$v = n/2$	$[\mu_1, \mu_2, \ldots, \mu_v]$
G_2		2	$[\gamma_1, \gamma_2]$
F_4		4	$[\gamma_1, \gamma_2, \gamma_3, \gamma_4]$
E_6		6	$[\gamma_1, \gamma_2, \gamma_3, \gamma_4, \gamma_5, \gamma_6]$
E_7		7	$[\gamma_1, \gamma_2, \gamma_3, \gamma_4, \gamma_5, \gamma_6, \gamma_7]$
E_8		8	$[\gamma_1, \gamma_2, \gamma_3, \gamma_4, \gamma_5, \gamma_6, \gamma_7, \gamma_8]$

$$x_i' = \sum_k a_{ik} x_k \; ; \; y_i' = \sum_k a_{ik} y_k \; ; \; (i, k = 1, \ldots, n). \tag{6.2}$$

Consider now the Kronecker product of the two vectors, $F_{ij} = x_i y_j$. This product has $n \times n$ components and transforms as a second rank tensor with respect to $GL(n)$

$$F_{ij}' = \sum_{k,l} a_{ik} a_{jl} F_{kl}. \tag{6.3}$$

Irreducible tensors with respect to $GL(n)$ are obtained by taking the symmetric and antisymmetric parts

$$S_{ik} = S_{ki} = \frac{1}{2} (F_{ik} + F_{ki}),$$

$$A_{ik} = -A_{ki} = \frac{1}{2} (F_{ik} - F_{ki}). \tag{6.4}$$

The symmetric tensor has $n(n+1)/2$ components, while the antisymmetric tensor has $n(n-1)/2$ components. In general, reducibility with respect to $GL(n)$ [and $gl(n)$] means to classify tensors according to their symmetry under interchange of indices. Young devised a procedure how to find the possible symmetry types of a tensor of rank t in a n dimensional space.

(i) Partition t into n integers

$$t = \lambda_1 + \ldots + \lambda_n \tag{6.5}$$

with

$$\lambda_1 \geq \lambda_2 \geq \ldots \geq \lambda_n \geq 0. \tag{6.6}$$

(ii) To each partition there correspond a graph (or tableau)

$$
\begin{array}{l}
\lambda_1 \ \square \ \square \ \ldots \ \square \\
\lambda_2 \ \square \ \ldots \ \square \\
\ \cdots \ \cdots \\
\lambda_n \ \square
\end{array}
\tag{6.7}
$$

The tensor is symmetric under interchange of the rows and antisymmetric under interchange of columns. The tableau is often denoted by $[\lambda_1, \ldots, \lambda_n]$ and zeros are deleted.

Example 1. Consider a second rank tensor $t = 2$ in a three dimensional space, $n = 3$.

The partitions are $[2], [1, 1]$ with Young diagrams

$$
\begin{array}{ll}
\square\square \equiv [2] & \text{Symmetric tensor} \\
\begin{array}{l}\square\\\square\end{array} \equiv [1, 1] & \text{Antisymmetric tensor}
\end{array}
\tag{6.8}
$$

6.3.2 Construction of Irreducible Tensors with Respect to GL(n). Young Method

Young devised a method to construct explicitly irreducible tensors with respect to $GL(n)$ starting from a generic tensor of rank r,

$$
F_{i_1 i_2 \ldots i_r} \ . \tag{6.9}
$$

To this end, we introduce the permutation group of n objects, S_n, with $n!$ elements written as

$$
\begin{pmatrix} 1 & 2 & \ldots & n \\ p_1 & p_2 & & p_n \end{pmatrix}. \tag{6.10}
$$

The simplest permutation is the transposition $1 \to 2, 2 \to 1$, which we write as (12),

$$
\begin{pmatrix} 1 & 2 \\ 2 & 1 \end{pmatrix} \equiv (12). \tag{6.11}
$$

Any permutation can be written as a product of transpositions. For example, the cyclic permutation (123) can be written as

$$\begin{pmatrix} 1\,2\,3 \\ 2\,3\,1 \end{pmatrix} = \begin{pmatrix} 1\,3 \\ 3\,1 \end{pmatrix} \begin{pmatrix} 1\,2 \\ 2\,1 \end{pmatrix} \quad \text{or} \quad (123) = (13)(12). \tag{6.12}$$

A permutation is called even if it is the product of an even number of transpositions, it is called odd if it is the product of an odd number of transpositions. We next construct the quantities

$$P = \sum_p p, \quad Q = \sum_q \delta_q q, \tag{6.13}$$

where δ_q is the parity of the permutation. P is called the "symmetrizer", Q the "antisymmetrizer", and

$$Y = QP \tag{6.14}$$

the Young operator.

To construct irreducible representations of $GL(n)$ with Young pattern

$$F \tag{6.15}$$

i_1	i_2	$\cdots$	i_{λ_1}
i_{λ_1+1}	$\cdots$	$i_{\lambda_1+\lambda_2}$	
$\cdots$	$\cdots$		
	$\cdots$	i_r	

in terms of tensors of rank r, we start from the general r-th rank tensor $F_{i_1 i_2 \ldots i_r}$ (6.9) and apply to it the Young operator $Y = QP$ where P is the operator for the horizontal permutations in the diagram and Q the operator for the vertical permutations.

Example 2. Construct the tensor

$$F \tag{6.16}$$

i_1	i_2
i_3	i_4

from the general tensor $F_{i_1 i_2 i_3 i_4}$.

The symmetrizer and anitsymmetrizer are

$$P = [e + (12)]\,[e + (34)],$$
$$Q = [e - (13)]\,[e - (24)]. \tag{6.17}$$

Applying them to F, we obtain

$$(PF)_{i_1 i_2 i_3 i_4} = F_{i_1 i_2 i_3 i_4} + F_{i_2 i_1 i_3 i_4} + F_{i_1 i_2 i_4 i_3} + F_{i_2 i_1 i_4 i_3}, \tag{6.18}$$

and

$$
\begin{aligned}
(QPF)_{i_1 i_2 i_3 i_4} = {} & F_{i_1 i_2 i_3 i_4} - F_{i_3 i_2 i_1 i_4} - F_{i_1 i_4 i_3 i_2} + F_{i_3 i_4 i_1 i_2} \\
& + F_{i_2 i_1 i_3 i_4} - F_{i_2 i_3 i_1 i_4} - F_{i_4 i_1 i_3 i_2} + F_{i_4 i_3 i_1 i_2} \\
& + F_{i_1 i_2 i_4 i_3} - F_{i_3 i_2 i_4 i_1} - F_{i_1 i_4 i_2 i_3} + F_{i_3 i_4 i_2 i_1} \\
& + F_{i_2 i_1 i_4 i_3} - F_{i_2 i_3 i_4 i_1} - F_{i_4 i_1 i_2 i_3} + F_{i_4 i_3 i_2 i_1}.
\end{aligned}
\tag{6.19}
$$

6.3.3   Irreducible Tensors with Respect to $SU(n)$

The irreducible representations of $GL(n)$ remain irreducible when we go to $U(n)$. If we go to the unimodular groups $SU(n)$, the representations corresponding to the patterns $[\lambda_1, \lambda_2, \ldots, \lambda_n]$ and $[\lambda_1 + s, \lambda_2 + s, \ldots, \lambda_n + s]$, s = integer, are equivalent. Thus, for $SU(n)$, we need to consider only patterns with one less row $[\lambda_1 - \lambda_n, \lambda_2 - \lambda_n, \ldots, \lambda_{n-1} - \lambda_n]$ obtained from the pattern $[\lambda_1, \lambda_2, \ldots, \lambda_n]$ by subtracting the last integer λ_n.

6.3.4   Irreducible Tensors with Respect to $SO(n)$

When we go from $U(n)$ to $SO(n)$, the representations in terms of tensors of a given symmetry are no longer irreducible. There is a new operation, called contraction, which commutes with orthogonal transformations. For $SO(n)$, the elements of the transformation matrix a_{ij} satisfy

$$
\sum_i a_{ij} a_{ik} = \delta_{jk}.
\tag{6.20}
$$

Contraction of a second rank tensor F_{ij} gives

$$
\tau = \sum_i F_{ii}.
\tag{6.21}
$$

In general, contraction of a rank-t tensor, gives a tensor of rank $t - 2$. Any tensor can be decomposed into a traceless part, plus the rest.

Example 3. Decomposition of a rank-2 tensor in n = 3 dimensions

$$GL(3) \qquad\qquad SO(3) \qquad\quad \text{dim } L$$

$$S_{ik} - \tfrac{1}{3}S_{ii}\delta_{ik} = \Sigma_{ik} \quad 5 \quad 2$$

$$S_{ik}$$

(6.22)

$$F_{ik}$$

$$\tfrac{1}{3}S_{ii}\delta_{ik} = \tau_{ik} \qquad 1 \quad 0$$

$$A_{ik} \qquad\qquad\qquad\qquad 3 \quad 1$$

The decomposition is thus

$$F_{ik} = A_{ik} + \Sigma_{ik} + \tau_{ik}. \tag{6.23}$$

The number of components of the tensors A, Σ, τ is shown in the column labeled dim.

6.4   Tensor Representations of Classical Compact Algebras

6.4.1   Unitary Algebras $u(n)$

Irreducible representations of $u(n)$ are characterized by n integers, satisfying the conditions

$$\lambda_1 \geq \lambda_2 \geq \ldots \geq \lambda_n \geq 0. \tag{6.24}$$

A graphical representation is provided by the Young tableau introduced previously for $U(n)$

$$\begin{array}{l} \lambda_1 \ \square\,\square\,\ldots\,\square \\ \lambda_2 \ \square\,\square\,\ldots\,\square \\ \qquad \ldots \\ \lambda_n \ \square \end{array}, \tag{6.25}$$

also written as $[\lambda_1, \lambda_2, \ldots, \lambda_n]$.

6.4.2   Special Unitary Algebras $su(n)$

Irreducible representations of $su(n)$ are characterized by $n - 1$ integers, satisfying the relations

$$\lambda_1 \geq \lambda_2 \geq \ldots \geq \lambda_{n-1} \geq 0. \tag{6.26}$$

Because of the special condition, S, some representations become equivalent.

Equivalence relation 1   *Start from the representations of u(n) and subtract the last integer* $[\lambda_1, \lambda_2, \ldots, \lambda_n] \equiv [\lambda_1 - \lambda_n, \lambda_2 - \lambda_n, \ldots, \lambda_{n-1} - \lambda_n, 0]$.

This equivalence relation was already quoted in Sect. 6.3.

Equivalence relation 2   *Start from the representations of u(n) and use* $[\lambda_1, \lambda_2, \ldots, \lambda_n] \equiv [\lambda_1 - \lambda_n, \lambda_1 - \lambda_{n-1}, \ldots, \lambda_1 - \lambda_2, 0]$.

This equivalence relation, when written at the level of $su(n)$ is $[\lambda_1, \lambda_2, \ldots, \lambda_n] \equiv [\lambda_1, \lambda_1 - \lambda_{n-1}, \ldots, \lambda_1 - \lambda_2]$ and is sometimes called particle-hole conjugation.

The equivalence relations 1 and 2 are used extensively.

Example 4. The first equivalence relation for su(3) gives

$$[4, 3, 1] \equiv [3, 2]. \tag{6.27}$$

Example 5. The second equivalence relation for su(3) gives

$$[3, 2] \equiv [3, 1]. \tag{6.28}$$

6.4.3 Orthogonal Algebras so(n), n = Odd

The irreducible representations are labeled by $v = (n - 1)/2$ integers, satisfying

$$\mu_1 \geq \mu_2 \geq \ldots \geq \mu_v \geq 0. \tag{6.29}$$

6.4.4 Orthogonal Algebras so(n), n = Even

The irreducible representations are labeled by $v = n/2$ integers, satisfying

$$\mu_1 \geq \mu_2 \geq \ldots \geq |\mu_v| \geq 0. \tag{6.30}$$

The last integer μ_v can be here positive, negative or zero. If $\mu_v \neq 0$, there are two irreducible representations, called mirror conjugate, with $\mu_v = \pm |\mu_v|$.

6.4.5 Symplectic Algebras sp(n), n = Even

The irreducible representations are labeled by $v = n/2$ integers, satisfying

$$\mu_1 \geq \mu_2 \geq \ldots \geq \mu_v \geq 0. \tag{6.31}$$

Example 6. Unitary algebras

$$\begin{aligned} u(2) \qquad & [\lambda_1, \lambda_2] \,, \; \lambda_1 \geq \lambda_2 \geq 0 \\ u(3) \;\; & [\lambda_1, \lambda_2, \lambda_3] \,, \;\; \lambda_1 \geq \lambda_2 \geq \lambda_3 \geq 0 \end{aligned} \tag{6.32}$$

Example 7. Special unitary algebras

$$\begin{aligned} su(2) \qquad\quad & [\lambda_1 - \lambda_2] = [f_1] \,, \; f_1 \geq 0 \\ su(3) \;\; & [\lambda_1 - \lambda_3, \lambda_2 - \lambda_3] \equiv [f_1, f_2] \,, \;\; f_1 \geq f_2 \geq 0 \end{aligned} \tag{6.33}$$

Example 8. Orthogonal algebras

$$\begin{aligned} so(2) \qquad & [\mu_1] \equiv M, \;\; |\,\mu_1\,| \geq 0 \\ so(3) \qquad & [\mu_1] \equiv L, \;\; \mu_1 \geq 0 \\ so(4) \;\; & [\mu_1, \mu_2] \equiv (\omega_1, \omega_2) \,; \; \mu_1 \geq |\,\mu_2\,| \geq 0 \\ so(5) \;\; & [\mu_1, \mu_2] \equiv (\tau_1, \tau_2) \,; \; \mu_1 \geq \mu_2 \geq 0 \end{aligned} \tag{6.34}$$

Often, in applications, the abstract labels are replaced by letters related to their physical interpretation, especially for orthogonal algebras and groups. In Example 8, the letter M (z-component of the angular momentum) is used to denote the representations of $so(2)$, and the letter L (angular momentum) to denote the representations of $so(3)$.

6.5   Spinor Representations

As discussed in Chap. 3, the group $SO(3)$ is doubly connected. It turns out that all orthogonal groups in odd number of dimensions $SO(2v + 1)$ are doubly connected, while the orthogonal groups in even number of dimensions, $SO(2v)$, are four-fold connected. As a result, orthogonal groups and algebras have another type of representations, called spinor representations, characterized by half-integer labels. (The additional two-fold connectedness of $SO(2v)$ produces mirror conjugate representations already discussed in the previous section) (Barut and Raçzka, 1986).

6.5.1   Orthogonal Algebras so(n), n = Odd

Spinor representations of $so(n), n = $ odd, are characterized by $v = (n-1)/2$ half-integers

$$\mu_1 \geq \mu_2 \geq \ldots \geq \mu_\nu \geq \frac{1}{2}. \tag{6.35}$$

6.5.2 Orthogonal Algebras $so(n)$, $n = Even$

Spinor representations of $so(n), n = $ odd, are characterized by $\nu = n/2$ half-integers

$$\mu_1 \geq \mu_2 \geq, \ldots \geq |\mu_\nu| \geq \frac{1}{2}. \tag{6.36}$$

As in the case of tensor representations, when $n = $ even, there are two irreducible representations, called mirror conjugate, with $\mu_\nu = \pm |\mu_\nu|$.

When spinor representations are included the algebras and groups are denoted $spin(n)$ and $Spin(n)$ respectively.

Example 9. Spinor representations

$$\begin{array}{ll} spin(2) & [\mu_1] \equiv M_J = \pm\frac{1}{2}, \pm\frac{3}{2}, \ldots \\ spin(3) & [\mu_1] \equiv J = \frac{1}{2}, \frac{3}{2}, \ldots \end{array} \tag{6.37}$$

6.6 Fundamental Representations

Any irreducible representation can be written as

$$[\lambda] = \sum_{i=1}^{l} f_i \, [\lambda^i], \tag{6.38}$$

where the f_i's are non-negative integers and the $[\lambda^i]$'s are called fundamental representations. Here $[\lambda]$ is a short-hand notation for the Young tableau characterizing the representation and $[\lambda^i]$ is a short-hand for the Young tableau characterizing the fundamental representations. The index i runs from 1 to the values given in Table 6.1 under "number of labels". The fundamental representations of the classical Lie algebras are listed below.

6.6.1 Unitary Algebras

There are n fundamental representations here

$$u(n)$$
$$[1, 0, 0, \ldots, 0]$$
$$[1, 1, 0, \ldots, 0] \, . \tag{6.39}$$
$$\ldots$$
$$[1, 1, 1, \ldots, 1]$$

6.6.2   Special Unitary Algebras

There are $n - 1$ fundamental representations

$$su(n)$$
$$[1, 0, 0, \ldots, 0]$$
$$[1, 1, 0, \ldots, 0] \, . \tag{6.40}$$
$$\ldots$$
$$[1, 1, 1, \ldots, 1]$$

6.6.3   Orthogonal Algebras, n = Odd

There are $v = (n - 1)/2$ fundamental representations, one of which is a spinor representation

$$spin(n), n = odd$$
$$[1, 0, \ldots, 0]$$
$$[1, 1, \ldots, 0]$$
$$\ldots \tag{6.41}$$
$$[1, 1, \ldots, 1, 0]$$
$$\left[\tfrac{1}{2}, \tfrac{1}{2}, \ldots, \tfrac{1}{2}, \tfrac{1}{2}\right]$$

6.6.4   Orthogonal Algebras, n = Even

There are $v = n/2$ fundamental representations, two of which are spinor representations

$$spin(n), n = even$$
$$[1, 0, \ldots, 0]$$
$$[1, 1, \ldots, 0]$$
$$\cdots \qquad (6.42)$$
$$[1, 1, \ldots 1, 0, 0]$$
$$\left[\tfrac{1}{2}, \tfrac{1}{2}, \ldots, \tfrac{1}{2}, \tfrac{1}{2}\right]$$
$$\left[\tfrac{1}{2}, \tfrac{1}{2}, \ldots, \tfrac{1}{2}, -\tfrac{1}{2}\right]$$

6.6.5 Symplectic Algebras

There are $v = n/2$ fundamental representations.

$$sp(n), n = even$$
$$[1, 0, \ldots, 0]$$
$$[1, 1, \ldots, 0] \qquad \cdot \qquad (6.43)$$
$$\cdots$$
$$[1, 1, \ldots, 1]$$

6.7 Realization of Bases

Bases can be realized in various ways. Three of them have been widely used.

1. Homogeneous polynomials
   The components of the basis are written as

$$x_1^{\lambda_1} x_2^{\lambda_2} \ldots \qquad (6.44)$$

   and the elements of the Lie algebra (2.37) act on them.
2. Column vectors
   The components of the basis are written as column vectors

$$\begin{pmatrix} \cdots \\ \cdots \\ \cdots \end{pmatrix} . \qquad (6.45)$$

   The elements of the Lie algebra (2.39) act on these column vectors.
3. Boson creation operators on a vacuum
   The components of the basis are written as

$$b_\alpha^\dagger b_{\alpha'}^\dagger \ldots |0\rangle . \qquad (6.46)$$

The elements of the Lie algebra (2.40) act on these states.

In addition to these realizations, others have been used for applications in physics. Two of these, important for applications to fermionic systems are:

4. Polynomials in Grassmann variables θ_i

These are used in connection with the realization (2.44) of the Lie algebra.

5. Fermion creation operators on a vacuum

The components of the basis are written as

$$a_i^\dagger a_{i'}^\dagger \ldots |0\rangle . \tag{6.47}$$

The elements of the Lie algebra (2.46) act on this basis.

6.8   Chains of Algebras

For applications, it is necessary to characterize uniquely the basis, in other words to provide a complete set of quantum numbers (labels). This is done by introducing a chain of algebras,

$$\left| \begin{array}{ccc} g & \supset & g' & \supset & g'' & \supset & \ldots \\ \downarrow & & \downarrow & & \downarrow & \\ [\lambda] & & [\lambda'] & & [\lambda''] \end{array} \right) . \tag{6.48}$$

Since in applications in quantum mechanics, the representations are interpreted as quantum mechanical states, a notation often used is that introduced by Dirac, called bra-ket notation. A ket is denoted by $|\rangle$ and a bra by $\langle|$.

A crucial problem of representation theory is to find the irreducible representations of an algebra g' contained in a given representation of g (often called the branching problem).

6.9   Canonical Chains

The branching problem was solved completely by Gel'fand and Cetlin in a series of articles in the 1950s, for a particular chain of algebras, called the canonical chain of unitary and orthogonal algebras (Gel'fand and Cetlin, 1950a,b).

6.9.1   Unitary Algebras

The canonical chain is

$$u(n) \supset u(n-1) \supset u(n-2) \supset \ldots \supset u(1). \tag{6.49}$$

The labels (quantum numbers) are conveniently arranged into a pattern called Gel'fand pattern

$$
\begin{array}{ccccc}
\lambda_{1,n} & \lambda_{2,n} & & \lambda_{n-1,n} & \lambda_{n,n} \\
& \lambda_{1,n-1} & \ldots \quad \ldots & \lambda_{n-1,n-1} & \\
& \ldots & & \ldots & \\
& \lambda_{1,2} & \lambda_{2,2} & & \\
& & \lambda_{1,1} & &
\end{array}
\tag{6.50}
$$

The entries in this pattern are $\lambda_{i,j}$ where i labels the entries in the Young tableau, and j the algebra in the chain. For example, $\lambda_{n,n}$ is the nth entry in the Young tableau of $u(n)$.

The solution to the branching problem is that the labels must satisfy the inequalities

$$\lambda_{1,n} \geq \lambda_{1,n-1} \geq \lambda_{2,n} \geq \ldots \geq \lambda_{n,n} \geq 0$$

$$\ldots$$

$$\lambda_{1,2} \geq \lambda_{1,1} \geq \lambda_{2,2}. \tag{6.51}$$

These inequalities hold for any two rows in (6.50) and are often called triangular inequalities.

Example 10. Representations of u(4)

The complete basis is labelled by

$$
\begin{array}{cccc}
\lambda_{1,4} & \lambda_{2,4} & \lambda_{3,4} & \lambda_{4,4} \\
& \lambda_{1,3} & \lambda_{2,3} & \lambda_{3,3} \\
& & \lambda_{1,2} & \lambda_{2,2} \\
& & & \lambda_{1,1}
\end{array}
\tag{6.52}
$$

In the applications discussed in Chap. 9, one needs the branching of the totally symmetric representation $[N]$ of $u(4)$ with ket $\mid N_4; N_3; N_2; N_1 \rangle$, $N_4 = N$, and pattern

$$
\begin{array}{cccc}
N_4 & 0 & 0 & 0 \\
& N_3 & 0 & 0 \\
& & N_2 & 0 \\
& & & N_1
\end{array}
\tag{6.53}
$$

Use of Gel'fand inequalities gives $N_3 = N_4, N_4 - 1, \ldots, 0$; $N_2 = N_3, N_3 - 1, \ldots, 0$; $N_1 = N_2, N_2 - 1, \ldots, 0$.

6.9.2   Orthogonal Algebras

The canonical chain is

$$so(n) \supset so(n-1) \supset so(n-2) \supset \ldots \supset so(2). \tag{6.54}$$

For $n = 2k + 2 = even$, the Gel'fand pattern is

$$
\begin{array}{cccccc}
\mu_{1,2k+1} & & \mu_{2,2k+1} & \cdots & \mu_{k,2k+1} & \mu_{k+1,2k+1} \\
& \mu_{1,2k} & & & \mu_{k,2k} & \\
& \mu_{1,2k-1} & & & \mu_{k,2k-1} & \\
& & \cdots & & \cdots & \\
& & \mu_{1,4} & \mu_{2,4} & & \\
& & \mu_{1,3} & \mu_{2,3} & & \\
& & & \mu_{1,2} & & \\
& & & \mu_{1,1} & &
\end{array}
\tag{6.55}
$$

There is an alternation between even and odd algebras, ending with $so(5) \supset so(4) \supset so(3) \supset so(2)$. The triangular inequalities are:

$$\mu_{1,2k+1} \geq \mu_{1,2k} \geq \mu_{2,2k+1} \geq \cdots \geq \mu_{k,2k} \geq | \, \mu_{k+1,2k+1} \, |$$

$$\mu_{1,2k} \geq \mu_{1,2k-1} \geq \mu_{2,2k} \geq \cdots \geq \mu_{k,2k} \geq | \, \mu_{k,2k-1} \, |$$

$$\mu_{1,2k-1} \geq \mu_{1,2k-2} \geq \cdots \geq \mu_{k-1,2k-2} \geq | \, \mu_{k,2k-1} \, | \, . \tag{6.56}$$

For $n = 2k + 1 = odd$, the Gel'fand pattern is

$$
\begin{array}{cccccc}
\mu_{1,2k} & & & & \mu_{k,2k} & \\
\mu_{1,2k-1} & & & & \mu_{k,2k-1} & \\
& \mu_{1,2k-2} & & & \mu_{k-1,2k-2} & \\
& \cdots & & & \cdots & \\
& & \mu_{1,4} & \mu_{2,4} & & \\
& & \mu_{1,3} & \mu_{2,3} & & \\
& & & \mu_{1,2} & & \\
& & & \mu_{1,1} & &
\end{array}
\tag{6.57}
$$

There is alternation between odd and even algebras, still ending with $so(5) \supset so(4) \supset so(3) \supset so(2)$ with inequalities

$$\mu_{1,2k} \geq \mu_{1,2k-1} \geq \cdots \geq \mu_{k,2k} \geq | \, \mu_{k,2k-1} \, |$$

$$\mu_{1,2k-1} \geq \mu_{1,2k-2} \geq \cdots \geq \mu_{k-1,2k-2} \geq | \, \mu_{k,2k-1} \, | \, . \tag{6.58}$$

Example 11. Branching of so(4)

For the ket

$$
\left| \begin{array}{ccc}
so(4) & \supset so(3) & \supset so(2) \\
\downarrow & \downarrow & \downarrow \\
[\mu_1, \mu_2] & J & M
\end{array} \right)
\tag{6.59}
$$

the Gel'fand pattern is

$$
\begin{array}{cc}
\mu_1 & \mu_2 \\
J & \\
M &
\end{array}
\tag{6.60}
$$

with branching

$$
\mu_1 \geq J \geq |\mu_2| \quad , \qquad +J \geq M \geq -J.
\tag{6.61}
$$

6.10 Isomorphisms of Spinor Algebras

Because of the isomorphisms discussed in Chap. 2, one has

$$
\begin{aligned}
spin(2) &\sim u(1) \\
spin(3) &\sim su(2) \\
spin(4) &\sim su(2) \oplus su(2) \\
spin(5) &\sim sp(4) \\
spin(6) &\sim su(4)
\end{aligned}
\tag{6.62}
$$

After *spin*(6), the spinor algebras are no longer isomorphic to other (non-orthogonal) classical Lie algebras.

It is of interest to find the relation between the quantum numbers labeling the representations of two isomorphic algebras.

Example 12. The case $spin(3) \sim su(2)$

$$
\begin{array}{cc}
spin(3) & \sim su(2) \\
\downarrow & \downarrow \\
J & \lambda
\end{array}
\tag{6.63}
$$

The relation in this case is simply $J = \frac{\lambda}{2}$,

$$su(2) : \overbrace{\square\square\ldots\square}^{\lambda} \quad \equiv \quad spin(3) : J = \frac{\lambda}{2} \tag{6.64}$$

Example 13. The case $spin(6) \sim su(4)$

$$
\begin{array}{ccc}
spin(6) & \sim & su(4) \\
\downarrow & & \downarrow \\
[\sigma_1, \sigma_2, \sigma_3] & & [\lambda_1, \lambda_2, \lambda_3] \\
\sigma_1 \geq \sigma_2 \geq |\sigma_3| \geq 0 & \lambda_1 \geq \lambda_2 \geq \lambda_3 \geq 0
\end{array}
\tag{6.65}
$$

The relation in this case is

$$\lambda_1 = \sigma_1 + \sigma_2 \quad , \quad \lambda_2 = \sigma_1 - \sigma_3 \quad , \quad \lambda_3 = \sigma_2 - \sigma_3 \quad . \tag{6.66}$$

The correspondence between the fundamental representations is

$$
\begin{array}{cc}
spin(6) & su(4) \\
\left[\tfrac{1}{2}, \tfrac{1}{2}, \tfrac{1}{2}\right] [1,0,0] & \square \\
 & \\
\left[\tfrac{1}{2}, \tfrac{1}{2}, -\tfrac{1}{2}\right] [1,1,1] & \begin{array}{c}\square\\\square\\\square\end{array} \\
 & \\
[1,0,0] \quad [1,1,0] & \begin{array}{c}\square\\\square\end{array}
\end{array}
\tag{6.67}
$$

6.11 Nomenclature for $u(n)$

In physics textbooks, the representations of $u(n)$ are often referred according to the particles they describe. The totally symmetric representations

$$\overbrace{\square\square\ldots\square}^{N} \quad [N,0,0,\ldots,0] = [N,\dot{0}] \equiv [N] \tag{6.68}$$

describe bosons and are often referred to as bosonic. The notation $\left[N,\dot{0}\right]$ is rarely used. The totally antisymmetric representations

$$\begin{array}{c}\square\\\square\\\cdots\\\square\end{array}\qquad [1,1,\ldots,1] \equiv [\dot{1}] \qquad\qquad (6.69)$$

describe fermions and are often referred to as fermionic. The representations with mixed symmetry

$$\begin{array}{c}\square\square\ \cdots\ \square\\\square\ \cdots\ \square\\\vdots\\\square\end{array}\qquad\qquad (6.70)$$

describe particles with internal degrees of freedom.

6.12 Dimensions of the Representations

The dimensions of the representations are often used to check the branching rules. A general formula was provided by Weyl and is called the Weyl formula.

6.12.1 Dimensions of the Representations of $u(n)$

The dimension of the representation $[\lambda] \equiv [\lambda_1, \lambda_2, \ldots, \lambda_n]$ of $u(n)$ is given by the formula

$$\dim[\lambda] = \prod_{i<j} \frac{(\ell_i - \ell_j)}{(\ell_i^0 - \ell_j^0)} \qquad \ell_j^0 = n - j \quad,\quad \ell_j = \lambda_j + n - j, \qquad (6.71)$$

with $i, j = 1, \ldots, n$.

Example 14. Dimensions of the representations of u(3)

For $u(3)$,

$$\dim[\lambda] = \frac{(\ell_1 - \ell_2)(\ell_1 - \ell_3)(\ell_2 - \ell_3)}{(\ell_1^0 - \ell_2^0)(\ell_1^0 - \ell_3^0)(\ell_2^0 - \ell_3^0)}. \qquad (6.72)$$

From (6.71)

$$\dim[\lambda_1, \lambda_2, \lambda_3] = \frac{1}{2}(\lambda_1 - \lambda_2 + 1)(\lambda_1 - \lambda_3 + 2)(\lambda_2 - \lambda_3 + 1). \qquad (6.73)$$

For example, the dimension of the representation $[2, 1, 0]$ is

$$\dim[2, 1, 0] = 8. \tag{6.74}$$

It has become customary in particle physics to denote the representation with its dimension. This notation is ambiguous, as often there are different representations with the same dimension. The Gel'fand patterns of the basis states of the representation $[2, 1]$ of $u(3)$ and $su(3)$ are

$$
\left|\begin{matrix} 2 & 1 & 0 \\ & 2 & 1 \\ & & 2 \end{matrix}\right\rangle
\left|\begin{matrix} 2 & 1 & 0 \\ & 2 & 1 \\ & & 1 \end{matrix}\right\rangle
$$

$$
\left|\begin{matrix} 2 & 1 & 0 \\ & 2 & 0 \\ & & 2 \end{matrix}\right\rangle
\left|\begin{matrix} 2 & 1 & 0 \\ & 2 & 0 \\ & & 1 \end{matrix}\right\rangle
\left|\begin{matrix} 2 & 1 & 0 \\ & 2 & 0 \\ & & 0 \end{matrix}\right\rangle
$$

$$
\left|\begin{matrix} 2 & 1 & 0 \\ & 1 & 1 \\ & & 1 \end{matrix}\right\rangle
\tag{6.75}
$$

$$
\left|\begin{matrix} 2 & 1 & 0 \\ & 1 & 0 \\ & & 1 \end{matrix}\right\rangle
\left|\begin{matrix} 2 & 1 & 0 \\ & 1 & 0 \\ & & 0 \end{matrix}\right\rangle
$$

In this application, the basis states correspond to particles. The eight states of the representation 8 for baryons correspond to particles called $p, n, \Sigma^+, \Sigma^0, \Sigma^-, \Lambda^0, \Xi^0, \Xi^-$.

6.12.2   Dimensions of the Representations of $su(n)$

The dimensions of the representations of $su(n)$ can be obtained from those of $u(n)$ by equivalences. Thus, for the representation $[\lambda_1, \lambda_2, \dots, \lambda_{n-1}]$, formula (6.71) applies with $\lambda_n = 0$.

6.12.3   Dimensions of the Representations of $A_n \equiv su(n + 1)$

An alternative formula for the dimensions of the representations of $su(n)$, which makes the connection with those of the orthogonal and symplectic algebras given below clear, can be written for the Cartan algebras $A_n \equiv su(n + 1)$. This formula can be simply derived from (6.71).

One first constructs the quantities

$$g_i = n - i + 1 \quad , \quad m_i = \lambda_i + g_i \quad (i = 1, \ldots, n). \tag{6.76}$$

The dimensions of the representations of A_n are

$$\dim [\lambda] = \prod_i \left(\frac{m_i}{g_i} \right) \prod_{i<j} \left(\frac{m_i - m_j}{g_i - g_j} \right). \tag{6.77}$$

6.12.4 Dimensions of the Representations of $B_n \equiv so(2n + 1)$

One constructs the quantities

$$g_i = n - i + \frac{1}{2} \quad , \quad m_i = \mu_i + g_i \quad (i = 1, \ldots, n). \tag{6.78}$$

The dimensions of the representations of B_n are

$$\dim [\mu] = \prod_i \left(\frac{m_i}{g_i} \right) \prod_{i<j} \left(\frac{m_i - m_j}{g_i - g_j} \right) \prod_{i<j} \left(\frac{m_i + m_j}{g_i + g_j} \right). \tag{6.79}$$

Example 15. Dimensions of the representations of $so(3)$ *and* $so(5)$

The dimensions of the representations of $so(3)$ and $so(5)$ are given by

$so(3) \qquad\qquad \dim[\mu_1] = (2\mu_1 + 1)$
$so(5) \quad \dim [\mu_1, \mu_2] = \frac{1}{6} (\mu_1 - \mu_2 + 1) (\mu_1 + \mu_2 + 2) (2\mu_1 + 3) (2\mu_2 + 1).$
$$\tag{6.80}$$

6.12.5 Dimensions of the Representations of $C_n \equiv sp(2n)$

Here

$$g_i = n - i + 1 \quad , \quad m_i = \mu_i + g_i \quad (i = 1, \ldots, n). \tag{6.81}$$

The dimensions of the representations of C_n are

$$\dim [\mu] = \prod_i \left(\frac{m_i}{g_i}\right) \prod_{i<j} \left(\frac{m_i - m_j}{g_i - g_j}\right) \prod_{i<j} \left(\frac{m_i + m_j}{g_i + g_j}\right). \tag{6.82}$$

Example 16. Dimension of the representations of $sp(4)$

The dimensions of the representations of $sp(4)$ are given by

$$sp(4) \quad \dim [\mu_1, \mu_2] = \tfrac{1}{6} (\mu_1 - \mu_2 + 1) (\mu_1 + \mu_2 + 3) (\mu_1 + 2) (\mu_2 + 1). \tag{6.83}$$

6.12.6 Dimensions of the Representations of $D_n \equiv so(2n)$

Here

$$g_i = n - 1 \quad , \quad m_i = \mu_i + g_i \quad (i = 1, \dots, n). \tag{6.84}$$

The dimensions of the representations of D_n are

$$\dim [\mu] = \prod_{i<j} \left(\frac{m_i - m_j}{g_i - g_j}\right) \prod_{i<j} \left(\frac{m_i + m_j}{g_i + g_j}\right). \tag{6.85}$$

Example 17. Dimension of the representations of $so(4)$

The dimensions of the representations of $so(4)$ are

$$so(4) \quad \dim [\mu_1, \mu_2] = (\mu_1 - \mu_2 + 1) (\mu_1 + \mu_2 + 1). \tag{6.86}$$

6.13 Action of the Elements of g on the Basis $\mathcal{B}$

The action of the elements of g on the basis is of great interest in physics. We consider here the case of $u(n)$. The elements of $u(n)$, when written in the double index notation of Chap. 2, satisfy the commutation relations

$$[E_{ij}, E_{kl}] = \delta_{jk} E_{il} - \delta_{il} E_{kj} \qquad i, j, k, l = 1, \dots, n. \tag{6.87}$$

Consider only $E_{k,k}$, $E_{k,k-1}$ and $E_{k-1,k}$, since the action of the others can be obtained from the commutators of these. The action is

$$E_{k,k} |\lambda\rangle = (r_k - r_{k-1}) |\lambda\rangle$$

$$E_{k,k-1} |\lambda\rangle = \sum_{j=1}^{k-1} a_{k-1}^j |\lambda'\rangle$$

$$E_{k-1,k} |\lambda\rangle = \sum_{j=1}^{k-1} b_{k-1}^j |\lambda''\rangle \tag{6.88}$$

where $E_{k,k}$ is called the diagonal element (or operator), $E_{k,k-1}$ the lowering element (or operator), $E_{k-1,k}$ the raising element (or operator). In this equation, $|\lambda\rangle$ denotes a generic Gel'fand pattern of $u(n)$. The coefficients r_k are given by

$$r_0 = 0 \quad , \quad r_k = \sum_{j=1}^{k} \lambda_{j,k} \quad , \quad k = 1,\ldots,n, \tag{6.89}$$

while the coefficients a and b are given by

$$a_{k-1}^j = \left[-\frac{\prod\limits_{i=1}^{k} (\lambda_{i,k} - \lambda_{j,k-1} - i + j + 1) \prod\limits_{i=1}^{k-2} (\lambda_{i,k-2} - \lambda_{j,k-1} - i + j)}{\prod\limits_{\substack{i \neq j \\ i=1}}^{k-1} (\lambda_{i,k-1} - \lambda_{j,k-1} - i + j + 1) \prod\limits_{\substack{i \neq j \\ i=1}}^{k-1} (\lambda_{i,k-1} - \lambda_{j,k-1} - i + j)} \right]^{1/2}$$

$$b_{k-1}^j = \left[-\frac{\prod\limits_{i=1}^{k} (\lambda_{i,k} - \lambda_{j,k-1} - i + j) \prod\limits_{i=1}^{k-2} (\lambda_{i,k-2} - \lambda_{j,k-1} - i + j - 1)}{\prod\limits_{\substack{i \neq j \\ i=1}}^{k-1} (\lambda_{i,k-1} - \lambda_{j,k-1} - i + j) \prod\limits_{\substack{i \neq j \\ i=1}}^{k-1} (\lambda_{i,k-1} - \lambda_{j,k-1} - i + j - 1)} \right]^{1/2}$$

$$\tag{6.90}$$

The representation $|\lambda'\rangle$ is obtained from $|\lambda\rangle$ by replacing $\lambda_{j,k-1}$ by $\lambda_{j,k-1} - 1$. (This is the reason why the operator $E_{k,k-1}$ is called lowering operator). $|\lambda''\rangle$ is obtained from $|\lambda\rangle$ by replacing $\lambda_{j,k-1}$ by $\lambda_{j,k-1} + 1$. ($E_{k-1,k}$ is called raising operator).

These results were obtained by Baird and Biederharn in the 1960s (Baird and Biederharn, 1963).

Example 18. The algebra u(2)

The states are

$$\left| \begin{array}{cc} \lambda_{1,2} & \lambda_{2,2} \\ & \lambda_{1,1} \end{array} \right\rangle \tag{6.91}$$

The algebra contains four elements $E_{1,1}, E_{1,2}, E_{2,1}, E_{2,2}$. The action of the elements on the basis is

$$E_{1,1} \begin{vmatrix} \lambda_{1,2} & & \lambda_{2,2} \\ & \lambda_{1,1} & \end{vmatrix} = \lambda_{1,1} \begin{vmatrix} \lambda_{1,2} & & \lambda_{2,2} \\ & \lambda_{1,1} & \end{vmatrix}$$

$$E_{2,2} \begin{vmatrix} \lambda_{1,2} & & \lambda_{2,2} \\ & \lambda_{1,1} & \end{vmatrix} = (\lambda_{1,2} + \lambda_{2,2} - \lambda_{1,1}) \begin{vmatrix} \lambda_{1,2} & & \lambda_{2,2} \\ & \lambda_{1,1} & \end{vmatrix}$$

$$E_{2,1} \begin{vmatrix} \lambda_{1,2} & & \lambda_{2,2} \\ & \lambda_{1,1} & \end{vmatrix} = [(\lambda_{1,2} - \lambda_{1,1} + 1)(\lambda_{1,1} - \lambda_{2,2})]^{1/2} \begin{vmatrix} \lambda_{1,2} & & \lambda_{2,2} \\ & \lambda_{1,1} - 1 & \end{vmatrix}$$

$$E_{1,2} \begin{vmatrix} \lambda_{1,2} & & \lambda_{2,2} \\ & \lambda_{1,1} & \end{vmatrix} = [(\lambda_{1,1} - \lambda_{1,2})(\lambda_{2,2} - \lambda_{1,1} - 1)]^{1/2} \begin{vmatrix} \lambda_{1,2} & & \lambda_{2,2} \\ & \lambda_{1,1} + 1 & \end{vmatrix}$$

$$\text{(6.92)}$$

Example 19. The algebra su(2)

In Gel'fand notation the basis states are

$$\begin{vmatrix} \lambda_{1,2} = 2J & & \lambda_{2,2} = 0 \\ & \lambda_{1,1} = M + J & \end{vmatrix}. \tag{6.93}$$

In the usual notation in quantum mechanics the basis states are written as $| J, M \rangle$. The action of the elements on the basis is

$$E_{1,1} | J, M \rangle = (M + J) | J, M \rangle$$
$$E_{2,2} | J, M \rangle = (-M + J) | J, M \rangle. \tag{6.94}$$

From these one obtains

$$\frac{1}{2} (E_{1,1} - E_{2,2}) | J, M \rangle = M | J, M \rangle. \tag{6.95}$$

The action of the raising and lowering operators is

$$E_{2,1} | J, M \rangle = [J(J + 1) - M(M - 1)]^{1/2} | J, M - 1 \rangle$$
$$E_{1,2} | J, M \rangle = [J(J + 1) - M(M + 1)]^{1/2} | J, M + 1 \rangle. \tag{6.96}$$

In quantum mechanics textbooks, the elements of the algebra are denoted by J_z, J_+, J_- with

$$J_z = \frac{1}{2} (E_{1,1} - E_{2,2})$$
$$J_- = E_{1,2}$$

$$J_+ = E_{2,1} \tag{6.97}$$

In this notation, the action of the elements of the algebra is

$$J_z |J, M\rangle = M |J, M\rangle$$
$$J_- | J, M \rangle = [J(J + 1) - M(M - 1)]^{1/2} | J, M - 1 \rangle$$
$$J_+ | J, M \rangle = [J(J + 1) - M(M + 1)]^{1/2} | J, M + 1 \rangle. \tag{6.98}$$

6.14   Tensor Products

With the representations $[\lambda_1, \lambda_2, \ldots, \lambda_n]$ one can form tensor products. The outer product of two tensors is denoted by

$$[\lambda_1', \lambda_2', \ldots, \lambda_n'] \otimes [\lambda_1'', \lambda_2'', \ldots, \lambda_n''] = \sum \oplus [\lambda_1, \lambda_2, \ldots, \lambda_n]. \tag{6.99}$$

A crucial problem is to find what are the representations contained in the product. For $u(n)$ and $su(n)$ the representations contained in the product can be simply obtained using a set of rules, known also as Young calculus.

Rule 1. *Product by a symmetric representation*

Consider the product of a generic representation $[\lambda_1', \lambda_2', \ldots, \lambda_n']$ by a symmetric representation $[\lambda_1'', 0, \ldots, 0]$, for example

$$\begin{array}{l} \square\,\square \otimes \square\,\square \\ \square \end{array} \tag{6.100}$$

Replace the second factor by a's

$$\begin{array}{l} \square\,\square \otimes a\,a \\ \square \end{array}. \tag{6.101}$$

Place the a's in all possible ways but no two a's in the same column

$$\begin{array}{l} \square\,\square\,a\,a \oplus \square\,\square\,a \oplus \square\,\square\,a \oplus \square\,\square \\ \square \qquad\quad \square\ a \qquad \square \qquad\quad \square\ a \\ \qquad\qquad\qquad\quad a \qquad\quad a \end{array} \tag{6.102}$$

Example 20. Product of fundamental representations

$$\square \otimes \square = \square\,\square \oplus \square \atop \square \tag{6.103}$$

When written in the notation (6.99)

$$[1] \otimes [1] = [2] \oplus [1,1].\tag{6.104}$$

Rule 2. *Product by a generic representation*

Consider the product of a generic representation $[\lambda_1', \lambda_2', \ldots, \lambda_n']$ by a generic representation $[\lambda_1'', \lambda_2'', \ldots, \lambda_n'']$, for example

$$\begin{array}{cc} \square\,\square & \otimes \;\; \square\,\square \\ \square & \square \end{array}\tag{6.105}$$

Replace the second factor by a's, b's, ...

$$\begin{array}{cc} \square\,\square & \otimes \; a\ a \\ \square & b \end{array}\tag{6.106}$$

Place the a's in all possible ways but no two a's in the same column. Place the b's in all possible ways but no two b's in the same column. The b's must form with the a's an admissible sequence when read from right to left (i.e. in inverse order $\ldots, c, b, a$) in the first row, then in the second row, ...

Definition 1. A sequence of letters a,b,c,... is admissible if at any point in the sequence at least as many a's have occurred as b's, at least as many b's have occurred as c's, etc.

Thus *abcd* and *aabcb* are admissible, while *baa* and *acb* are not. To implement this ordering rule, start from the first row and take the boxes from right to left. Then, run through the second row from right to left and so on. This ordered sequence contains boxes with labels and empty boxes. The sequence of letters must be admissible.

Example 21. Tensor product of representations of $u(3)$

Consider the product $[2, 1] \otimes [1, 1]$. Using rule 2 we have

$$\begin{array}{lcccc}
\square\square & \otimes\; a & =\; \square\square ab & \oplus\; \square\square a & \oplus\; \square\square a & \oplus\; \square\square b \\
\square & b & \square & \square b & \square & \square a \\
& & * & & b & *
\end{array}$$

$$\begin{array}{ccc}
\oplus\; \square\square & \oplus\; \square\square b & \oplus\; \square\square \\
\square a & \square & \square b \\
b & a & a \\
& * & *
\end{array}\tag{6.107}$$

The admissible and not admissible * sequences are

$$
\begin{array}{ll}
b\ a\ \square\ \square\ \square\ * & \text{not admissible} \\
a\ \square\ \square\ b\ \square & \text{admissible} \\
a\ \square\ \square\ \square\ b & \text{admissible} \\
b\ \square\ \square\ a\ \square\ * & \text{not admissible}\,. \\
\square\ \square\ a\ \square\ b & \text{admissible} \\
b\ \square\ \square\ \square\ a\ * & \text{not admissible} \\
\square\ \square\ b\ \square\ a\ * & \text{not admissible}
\end{array}
\tag{6.108}
$$

(Not admissible sequences are identified in (6.107) by a star * placed under the corresponding Young diagram). Therefore, deleting the not admissible * sequences, one has

$$
u(3): \quad [2,1] \otimes [1,1] = [3,2] \oplus [3,1,1] \oplus [2,2,1]. \tag{6.109}
$$

Rule 3. *For su(n) use equivalences when necessary*

Example 22. Tensor product of representations of $su(3)$

From (6.109), one has

$$
su(3): \quad (2,1) \otimes (1,1) = (3,2) \oplus (2,0) \oplus (1,1). \tag{6.110}
$$

Parentheses () have been used to denote representations of $su(3)$ to distinguish them from those of $u(3)$ denoted by brackets [].

Rule 4. *Check dimensions if needed*

The dimensions of the representations in the product above are

$$
su(3): \quad 8 \otimes \bar{3} = 15 \oplus 6 \oplus \bar{3}. \tag{6.111}
$$

Note that the two representations $(1,0)$ and $(1,1)$ have the same dimension, 3. A bar is usually placed above the dimension of the second representation, when this situation occurs.

Example 23. Another tensor product of representations of $u(3)$

Consider the product $[2,1] \otimes [2,1]$. Using rule 2,

$$\square\square \otimes \begin{matrix} a\,a \\ b \end{matrix} = \begin{matrix}\square\square\,a\,a\,b\\\square\\ *\end{matrix} \oplus \begin{matrix}\square\square\,a\,a\\\square\end{matrix} \oplus \begin{matrix}\square\square\,a\,a\\\square\,b\\ b\end{matrix} \oplus \begin{matrix}\square\square\,a\,b\\\square\,a\\ *\end{matrix}$$

$$\oplus \begin{matrix}\square\square\,a\\\square\,a\,b\end{matrix} \oplus \begin{matrix}\square\square\,a\\\square\,a\\ b\\ *\end{matrix} \oplus \begin{matrix}\square\square\,a\,b\\\square\\ a\end{matrix} \oplus \begin{matrix}\square\square\,a\\\square\,b\\ a\end{matrix} \qquad . \qquad (6.112)$$

$$\oplus \begin{matrix}\square\square\,b\\\square\,a\\ a\\ *\end{matrix} \oplus \begin{matrix}\square\square\\\square\,a\\ a\,b\end{matrix}$$

The not admissible * sequences are

$$\begin{matrix} b\,a\,a\,\square\,\square\,\square\,* \\ b\,a\,\square\,\square\,a\,\square\,* \\ b\,a\,\square\,\square\,\square\,a\,* \\ b\,\square\,\square\,a\,\square\,a\,* \end{matrix} \qquad (6.113)$$

from which we obtain

$$u(3): \quad [2,1] \otimes [2,1] = [4,2] \oplus [4,1,1] \oplus [3,3] \oplus [3,2,1] \oplus [3,2,1] \oplus [2,2,2]$$
$$su(3): \quad (2,1) \otimes (2,1) = (4,2) \oplus (3,0) \oplus (3,3) \oplus (2,1) \oplus (2,1) \oplus (0,0)$$
$$(6.114)$$

The dimensions of the representations in the product above are

$$su(3): \quad 8 \otimes 8 = 27 \oplus 10 \oplus \overline{10} \oplus 8 \oplus 8 \oplus 1. \qquad (6.115)$$

Multiplication rules for $so(n)$ and $sp(n)$ are rather complicated. Simple rules can be obtained only by using the isomorphisms discussed in Chap. 2.

Example 24. Multiplication rules for so(3)

Consider, for example, the product $(1) \otimes (1)$. Using the relation $\lambda = 2J$, this product can be converted to

$$su(2): \quad [2] \otimes [2]. \qquad (6.116)$$

This product can be performed using the rules 1 and 3,

$$\square\square \otimes aa = \square\square\, a\, a \,\oplus\, \begin{array}{c}\square\square\, a \\ a\end{array} \,\oplus\, \begin{array}{c}\square\square \\ a\ a\end{array} \tag{6.117}$$

that is

$$[2] \otimes [2] = [4] \oplus [2] \oplus [0]. \tag{6.118}$$

This can be converted back to $so(3)$ using $J = \lambda/2$, with result

$$(1) \otimes (1) = (2) \oplus (1) \oplus (0). \tag{6.119}$$

Example 25. Multiplication rules of so(6)

Consider as another example, the product $(2,0,0) \otimes (1,0,0)$. Using the relations $\lambda_1 = \sigma_1 + \sigma_2, \lambda_2 = \sigma_1 - \sigma_3, \lambda_3 = \sigma_2 - \sigma_3$ given in (6.66), this product can be converted to

$$su(4): \qquad [2,2,0] \otimes [1,1,0]. \tag{6.120}$$

This product can be performed using the rules for $su(4)$

$$\begin{array}{c}\square\square \\ \square\square\end{array} \otimes \begin{array}{c}a \\ b\end{array} = \begin{array}{c}\square\square\, a \\ \square\square\, b\end{array} \oplus \begin{array}{c}\square\square\, a \\ \square\square \\ b\end{array} \oplus \begin{array}{c}\square\square \\ \square\square \\ a \\ b\end{array} \tag{6.121}$$

that is

$$[2,2,0] \otimes [1,1,0] = [3,3,0] \oplus [3,2,1] \oplus [1,0,0]. \tag{6.122}$$

This formula can now be converted back to $so(6)$ using $\sigma_1 = \frac{\lambda_1+\lambda_2-\lambda_3}{2}, \sigma_2 = \frac{\lambda_1-\lambda_2+\lambda_3}{2}, \sigma_3 = \frac{\lambda_1-\lambda_2-\lambda_3}{2}$, with result

$$(2,0,0) \otimes (1,0,0) = (3,0,0) \oplus (2,1,0) \oplus (1,0,0). \tag{6.123}$$

6.15 Non-canonical Chains

Gel'fand canonical chains of unitary and orthogonal algebras provide a complete solution to the branching problem for these algebras. However, in most problems in physics, one needs to consider non-canonical chains. For example, quite often the rotation algebra $so(3)$ needs to be considered as a subalgebra of $u(n)$. Also, the branching problem for symplectic algebras $sp(n)$, not addressed by Gel'fand and Cetlin, needs be considered.

Example 26. The chain $u(6) \supset so(6) \supset so(5) \supset so(3) \supset so(2)$

This chain has two non-canonical steps, $u(6) \supset so(6)$ and $so(5) \supset so(3)$ and illustrates two of the most important cases often encountered: $u(n) \supset o(n); u(n) \supset sp(n); o(n) \supset o(n-2); sp(n) \supset sp(n-2)$. The other two cases, $u(n) \supset sp(n)$ and $sp(n) \supset sp(n-2)$, are also encountered, especially the former, as discussed in Chap. 10.

The decomposition of irreducible representations of g into representations of g' for non-canonical chains

$$\left| \begin{matrix} g & \supset & g' \\ \downarrow & & \downarrow \\ [\lambda] & & [\mu] \end{matrix} \right) \tag{6.124}$$

is one of the most difficult problems in group theory. A method often used is the so-called building-up process. The decomposition is constructed by taking successive products of the fundamental representations of g and g'. If one of the algebras is $su(n)$, equivalences are used. The method is illustrated by considering a problem of interest in nuclear physics.

Example 27. Decomposition of representations of $u(3)$ and $su(3)$ into representations of $so(3)$

In order to decompose representations of $u(3)$ and $su(3)$ into representations of $so(3)$ it is convenient to consider the decomposition $u(3) \supset so(3)$ and use equivalence relations to obtain $su(3) \supset so(3)$. It is also convenient, for clarity, to use Young notation to label representations of $u(3)$ and $su(3)$ and angular momentum notation to label representations of $so(3)$. In this notation, the fundamental representation $[1]$ of $u(3)$ has Young tableau $\square$, while the fundamental representation (1) of $so(3)$ has $L = 1$. The representation $\square$ of $u(3)$ contains only the representation (1) of $so(3)$, as one can see by noting that $\dim[\square] = 3 = \dim(1)$ and that there are no other representations of $so(3)$ with $\dim = 3$. (The dimensions of the representations play an important role in the building-up process. The dimension of the representations of $u(3)$ is given in (6.71), while that of $so(3)$ is given in (6.79).)The decomposition of the fundamental representation of $u(3)$ into representations of $so(3)$ is thus

$$\left| \begin{matrix} u(3) & \supset & so(3) \\ \downarrow & & \downarrow \\ [1] \equiv \square & & (1) \end{matrix} \right) . \tag{6.125}$$

Consider now the products of representations of $u(3)$ obtained by using the rules of Sect. 6.14 and of $so(3)$ obtained either by using the rules of Sect. 6.14 or, equivalently, by using the angular momentum rule

$$| L_1 - L_2 | \leq L \leq | L_1 + L_2 | \tag{6.126}$$

which results from the isomorphism $so(3) \sim su(2)$, as described in Example 24. The product of the fundamental representation $[1] \equiv \square$ of $u(3)$ with itself, and of the fundamental representation (1) of $so(3)$ with itself gives

$$u(3): \quad \square \otimes \square = \square\square \oplus \begin{matrix}\square\\\square\end{matrix}$$

$$so(3): \quad (1) \otimes (1) = (0) \oplus (1) \oplus (2). \tag{6.127}$$

The representation $[1, 1]$ of $u(3)$ contains only the representation (1) of $so(3)$ as one can see by noting that $\dim[1, 1] = 3 = \dim(1)$ and that there is no other representation of $so(3)$ with $\dim = 3$. Thus

$$\left| \begin{matrix} u(3) \supset so(3) \\ \downarrow \qquad \downarrow \\ [1, 1] \qquad (1) \end{matrix} \right). \tag{6.128}$$

The remaining representations $(0) \oplus (2)$ of $so(3)$ must belong to the representation $\square\square \equiv [2]$ of $u(3)$. This is verified by a dimensional check, $\dim[2] = 6$, $\dim((0) \oplus (2)) = 1 + 5 = 6$. Thus

$$\left| \begin{matrix} u(3) \supset \quad so(3) \\ \downarrow \qquad \downarrow \\ [2] \qquad (0) \oplus (2) \end{matrix} \right). \tag{6.129}$$

When going from $u(3)$ to $su(3)$, by virtue of equivalence relation 2, we have $[1, 1] \equiv [1]$, again showing that the representation $[1]$ of $su(3)$ contains only (1) of $so(3)$.

Next, consider the products

$$u(3): \quad \begin{matrix}\square\\\square\end{matrix} \otimes \square = \begin{matrix}\square\square\\\square\end{matrix} \oplus \begin{matrix}\square\\\square\\\square\end{matrix}$$

$$so(3): \quad (1) \otimes (1) = (0) \oplus (1) \oplus (2). \tag{6.130}$$

The dimension of the representation $[1, 1, 1]$ of $u(3)$ is 1 and therefore

$$\left| \begin{matrix} u(3) \supset so(3) \\ \downarrow \qquad \downarrow \\ [1, 1, 1] \qquad (0) \end{matrix} \right). \tag{6.131}$$

Consequently $[2, 1]$ must contain the remaining representations $(1) \oplus (2)$

$$\left.\begin{matrix} u(3) & \supset & so(3) \\ \downarrow & & \downarrow \\ [2,1] & & (1) \oplus (2) \end{matrix}\right). \tag{6.132}$$

The dimensional check is $\dim[2,1] = 8$, $\dim((1) \oplus (2)) = 3+5 = 8$. When going from $u(3)$ to $su(3)$, use the equivalence relation 1 gives

$$\begin{matrix} \square \\ \square \\ \square \end{matrix} \equiv [1,1,1] \equiv [0]. \tag{6.133}$$

A tedious but straightforward procedure, sometimes called plethism, gives then Table 6.2. In this table, for clarity, both the representations of $u(3)$ and of $su(3)$ (with their equivalences) are given. Also the number t, the rank of the tensor in (8.5), is given. This number is important in applications since it denotes the total number of particles

Table 6.2 Decomposition of representations of $u(3)$ and $su(3)$ into representations of $so(3)$

t	$u(3)$	$su(3)$	$so(3)$
0	[0]	[0]	0
1	[1]	[1]	1
2	[2]	[2]	0, 2
	[1, 1]	[1, 1] ≡ [1]	1
3	[3]	[3]	1, 3
	[2, 1]	[2, 1]	1, 2
	[1, 1, 1]	[1, 1, 1] ≡ [0]	0
4	[4]	[4]	0, 2, 4
	[3, 1]	[3, 1]	1, 2, 3
	[2, 2]	[2, 2] ≡ [2]	0, 2
	[2, 1, 1]	[2, 1, 1] ≡ [1]	1
5	[5]	[5]	1, 3, 5
	[4, 1]	[4, 1]	1, 2, 3, 4
	[3, 2]	[3, 2] ≡ [3, 1]	1, 2, 3
	[3, 1, 1]	[3, 1, 1] ≡ [2]	0, 2
	[2, 2, 1]	[2, 2, 1] ≡ [1, 1]	1
6	[6]	[6]	0, 2, 4, 6
	[5, 1]	[5, 1]	1, 2, 3, 4, 5
	[4, 2]	[4, 2]	$0, 2^2, 3, 4$
	[4, 1, 1]	[4, 1, 1] ≡ [3]	1, 3
	[3, 3]	[3, 3] ≡ [3]	1, 3
	[3, 2, 1]	[3, 2, 1] ≡ [2, 1]	1, 2
	[2, 2, 2]	[2, 2, 2] ≡ [0]	0

$$t = \lambda_1 + \ldots + \lambda_n. \tag{6.134}$$

Consider now the representation $[4, 2]$ of $su(3)$. From the table, one can see that it contains two $L = 2$ representations of $so(3)$. Thus the decomposition is not unique and one needs an additional quantum number to distinguish the two $L = 2$ representations. The identification and use of this quantum number (missing label) is one of the most subtle points of representation theory. To find how many missing labels there are in a given problem, go to the canonical chain (which has no missing labels) and count the total number of labels.

Example 28. Missing labels for $su(3) \supset so(3)$

Consider the non-canonical chain

$$\left| \begin{array}{ccc} su(3) & \supset so(3) \supset so(2) \\ \downarrow & \downarrow \quad\quad \downarrow \\ [\lambda_1, \lambda_2] & L \quad\quad M \end{array} \right) \tag{6.135}$$

and its associated canonical chain

$$\left| \begin{array}{ccc} su(3) & \supset & u(2) & \supset u(1) \\ \downarrow & & \downarrow & \downarrow \\ [\lambda_{1,3}, \lambda_{2,3}, 0] & & [\lambda_{1,2}, \lambda_{2,3}] & [\lambda_{1,1}] \end{array} \right). \tag{6.136}$$

The non-canonical chain has four labels, while the canonical chain has five. There is thus one missing label in the non-canonical chain.

Extensive tables for the non-canonical reductions

$$\begin{cases} u(7) \supset so(7) \\ u(5) \supset so(5) \\ u(3) \supset so(3) \end{cases} \begin{cases} u(8) \supset sp(8) \\ u(6) \supset sp(6) \\ u(4) \supset sp(4) \end{cases}$$

$$\begin{cases} u(6) \supset so(6) \\ u(4) \supset so(4) \end{cases} \tag{6.137}$$

exist (Hamermesh, 1962). Tables are also available for the reduction $so(5) \supset so(3)$ Iachello and Arima (1987).

Although not of the type mentioned at the beginning of the section, it is worth noting that tables of non-canonical reductions $sp(8) \supset spin(3), sp(6) \supset spin(3), sp(4) \supset spin(3)$ are also available (Flowers, 1952).

Chapter 7
Casimir Operators and Their Eigenvalues

7.1 Definitions

An operator which commutes with all the elements of a Lie algebra, g, is called an invariant, or Casimir operator, C

$$[C, X_\rho] = 0 \qquad \text{for any } X_\rho \in g \quad . \tag{7.1}$$

The operator is called of order p, if it is built from products of p elements

$$C_p = \sum_{\alpha_1, \alpha_2, \ldots, \alpha_p} f^{\alpha_1 \alpha_2 \ldots \alpha_p} X_{\alpha_1} X_{\alpha_2} \ldots X_{\alpha_p}. \tag{7.2}$$

It lies in the enveloping algebra of g, $T(g)$.

7.2 Independent Casimir Operators

The number of independent Casimir operators, C, of a Lie algebra g, is equal to the rank l of g, and hence equal to the number of labels that characterize the irreducible representations of g. As mentioned in Chap. 1, if C is a Casimir operator, so is $aC, C^2, \ldots$

7.2.1 Casimir Operators of $u(n)$

The algebra of $u(n)$ has n independent Casimir operators of order $1, 2, \ldots, n$

$$C_1, C_2, \ldots, C_n. \tag{7.3}$$

© Springer-Verlag Berlin Heidelberg 2015
F. Iachello, *Lie Algebras and Applications*, Lecture Notes in Physics 891,
DOI 10.1007/978-3-662-44494-8_7

A construction of these operators was given in Chap. 1 in terms of the structure constants $c^{\gamma}_{\alpha\beta}$. For the algebra $u(n)$, if the elements are denoted by $E_{ij}(i, j = 1, \ldots, n)$, that is if the algebra is realized as in Chap. 2, Sect. 2.17, the Casimir operators of order p can be written as

$$C_p = E_{i_1 i_2} E_{i_2 i_3} \ldots E_{i_{p-1} i_p} E_{i_p i_1} \quad p = 1, 2, \ldots, n \tag{7.4}$$

(summation over repeated indices). In particular

$$C_1 = E_{i_1 i_1}, \tag{7.5}$$

or, displaying explicitly the summation

$$C_1 = \sum_{i=1}^{n} E_{ii}. \tag{7.6}$$

Using the form (7.4), one can show that the Casimir operators so defined are independent (Barut and Rączka 1986).

7.2.2   Casimir Operators of $su(n)$

The independent Casimir operators of $su(n)$ are of order $2, 3, \ldots, n$

$$C_2, C_3, \ldots, C_n \tag{7.7}$$

that is the same as $u(n)$ but with C_1 omitted. The elements of $su(n)$ are obtained from those of $u(n)$ by keeping the off-diagonal elements the same, replacing the diagonal ones with

$$\tilde{E}_{ii} = E_{ii} - \frac{1}{n} \sum_{j=1}^{n} E_{jj}. \tag{7.8}$$

and deleting $\tilde{E}_{nn}$. The linear Casimir operator of $su(n)$ is

$$C_1 = \sum_{i=1}^{n} \tilde{E}_{ii} = 0, \tag{7.9}$$

and can thus be omitted.

7.2.3 Casimir Operators of so(n), n = Odd

The independent Casimir operators of $so(n), n =$ odd are

$$C_2, C_4, C_6, \ldots, C_{2v-2}, C_{2v}; \quad v = \frac{n-1}{2}. \tag{7.10}$$

These operators are all of even order.

7.2.4 Casimir Operators of so(n), n = Even

The independent Casimir operators of $so(n), n =$ even are

$$C_2, C_4, C_6, \ldots, C_{2v-2}, C_v'; \quad v = \frac{n}{2}. \tag{7.11}$$

The operators C are of even order, while C' is of even or odd order depending on whether v is even or odd. Comparing with the previous case (7.10) one can see that there is a peculiarity here, since the operator of order $2v$ is replaced by an operator of order v. The operator C_v' is needed to distinguish between the mirror conjugate representations $[\lambda_1, \lambda_2, \ldots, +\lambda_v]$ and $[\lambda_1, \lambda_2, \ldots, -\lambda_v]$.

Example 1. Casimir operators of $so(4)$ and $so(6)$

The Casimir operators of $so(4)$ are of order

$$C_2, C_2'. \tag{7.12}$$

The Casimir operators of $so(6)$ are of order

$$C_2, C_4, C_3'. \tag{7.13}$$

7.2.5 Casimir Operators of sp(n), n = Even

The independent Casimir operators of $sp(n), n = even$ are

$$C_2, C_4, C_6, \ldots, C_{2v-2}, C_{2v}; \quad v = \frac{n}{2}. \tag{7.14}$$

These operators are all of even order.

7.2.6 Casimir Operators of the Exceptional Algebras

The independent Casimir operators of the exceptional algebras are of order

$$
\begin{array}{ll}
G_2 & C_2, C_6 \\
F_4 & C_2, C_6, C_8, C_{12} \\
E_6 & C_2, C_5, C_6, C_8, C_9, C_{12} \\
E_7 & C_2, C_6, C_8, C_{10}, C_{12}, C_{14}, C_{18} \\
E_8 & C_2, C_8, C_{12}, C_{14}, C_{18}, C_{20}, C_{24}, C_{30}
\end{array}
\tag{7.15}
$$

7.3 Complete set of Commuting Operators

For any given Lie algebra g, one is often interested in constructing a complete set of commuting operators. This is done by considering the decomposition into subalgebras $g \supset g' \supset g'' \supset \dots$. In view of problems connected with missing labels, the construction of a complete set of commuting operators is straightforward only for canonical chains. The construction parallels that of the labels given in Chap. 6 (Gel'fand construction) and applies to the canonical chains $u(n) \supset u(n-1) \supset \dots \supset u(2) \supset u(1)$ and $so(n) \supset so(n-1) \supset \dots \supset so(2)$. The construction for the symplectic algebras $sp(n)$ is more complex and it will not be discussed here.

7.3.1 The Unitary Algebra $u(n)$

The commuting Casimir operators can be arranged into a triangular pattern

$$
\begin{array}{cccccc}
C_{1,n} & C_{2,n} & & C_{n-1,n} & & C_{n,n} \\
& C_{1,n-1} & & & C_{n-1,n-1} & \\
& \dots & & \dots & & \\
& & C_{1,2} & C_{2,2} & & \\
& & & C_{1,1} & &
\end{array}
\tag{7.16}
$$

where $C_{1,n}$ denotes $C_1(u(n)), \dots$.

7.3.2 The Orthogonal Algebra so(n), n = Odd

The complete set of commuting operators is, for $n = 2k + 1$,

$$
\begin{array}{ccc}
C_{2,2k+1} \; C_{4,2k+1} & & C_{2k-2,2k+1} \; C_{2k,2k+1} \\
C_{2,2k} \quad C_{4,2k} & & C_{2k-2,2k} \quad C'_{k,2k} \\
\cdots & \cdots & \\
& C_{2,4} \quad C'_{2,4} & \\
& C_{2,3} & \\
& C'_{1,2} &
\end{array}
\tag{7.17}
$$

7.3.3 The Orthogonal Algebra so(n), n = Even

The complete set of commuting operators is for $n = 2k + 2$,

$$
\begin{array}{cccc}
C_{2,2k+2} & C_{4,2k+2} & C_{2k,2k+2} & C'_{k+1,2k+2} \\
C_{2,2k+1} & & C_{2k,2k+1} & \\
\cdots & & \cdots & \\
& C_{2,4} \quad C'_{2,4} & & \\
& C_{2,3} & & \\
& C'_{1,2} & &
\end{array}
\tag{7.18}
$$

7.4 Eigenvalues of Casimir Operators

The eigenvalues of the Casimir operators of all classical Lie algebras in a representation $\mid \lambda \rangle$, denoted in short by

$$
\langle \lambda_1, \ldots, \lambda_n \mid C_p \mid \lambda_1, \ldots, \lambda_n \rangle = \langle C_p \rangle
\tag{7.19}
$$

were worked out in a series of papers by Perelomov and Popov in the 1960s (Perelomov and Popov 1966a,b). The algorithm to obtain the eigenvalues is described below.

7.4.1 The Algebras u(n) and su(n)

To find the eigenvalues of the Casimir operators, C_p, construct the quantities

$$
S_k = \sum_{i=1}^{n} (\ell_i^k - \rho_i^k) \,, \quad \rho_i = n - i \,, \quad \ell_i = m_i + n - i
\tag{7.20}
$$

with

$$m_i = \begin{cases} \lambda_i & \text{for } u(n) \\ \lambda_i - \frac{\lambda}{n} & \text{for } su(n) \end{cases}, \qquad \lambda = \sum_{i=1}^{n} \lambda_i. \qquad (7.21)$$

Construct the function

$$\varphi(z) = \sum_{k=2}^{\infty} a_k z^k, \qquad a_k = \sum_{j=1}^{k-1} \frac{(k-1)!}{j!(k-j)!} S_j. \qquad (7.22)$$

Define the quantities B_p by

$$\exp\{-\varphi(z)\} = 1 - \sum_{p=1}^{\infty} B_p z^{p+1}, \qquad B_0 = 0. \qquad (7.23)$$

Then, the eigenvalue of C_p in the representation $[\lambda_1, \lambda_2, \ldots, \lambda_n]$ is $\langle C_p \rangle = B_p - n B_{p-1}$. This algorithm gives the following eigenvalues of Casimir operators of order, $p \leq 3$. For $u(n)$

$$\langle C_1 \rangle = S_1$$
$$\langle C_2 \rangle = S_2 - (n-1)S_1$$
$$\langle C_3 \rangle = S_3 - \left(n - \frac{3}{2}\right) S_2 - \frac{1}{2} S_1^2 - (n-1)S_1$$

$$\cdots \qquad\qquad (7.24)$$

and for $su(n)$

$$\langle C_1 \rangle = 0$$
$$\langle C_2 \rangle = S_2$$
$$\langle C_3 \rangle = S_3 - \left(n - \frac{3}{2}\right) S_2$$

$$\cdots \qquad\qquad (7.25)$$

Example 2. Eigenvalue of the quadratic Casimir operator of su(2) in the representation $[\lambda_1]$

From

$$S_2 = \sum_{i=1}^{2} (\ell_i^2 - \rho_i^2)$$

$$\rho_1 = 1, \ \rho_2 = 0$$

$$\ell_1 = m_1 + 1, \ \ell_2 = m_2$$

$$m_1 = \lambda_1 - \frac{\lambda_1}{2}, \ m_2 = -\frac{\lambda_1}{2} \tag{7.26}$$

one obtains

$$\langle C_2 \rangle = \frac{1}{2}\lambda_1(\lambda_1 + 2). \tag{7.27}$$

Example 3. Eigenvalue of the quadratic Casimir operator of su(3) in the representation $[\lambda_1, \lambda_2]$

From

$$S_2 = \sum_{i=1}^{3} (\ell_i^2 - \rho_i^2)$$

$$\rho_1 = 2, \ \rho_2 = 1, \ \rho_3 = 0$$

$$\ell_1 = m_1 + 2, \ \ell_2 = m_2 + 1, \ \ell_3 = m_3$$

$$m_1 = \lambda_1 - \frac{\lambda_1 + \lambda_2}{3}, \ m_2 = \lambda_2 - \frac{\lambda_1 + \lambda_2}{3}, \ m_3 = -\frac{\lambda_1 + \lambda_2}{3} \tag{7.28}$$

one obtains

$$\langle C_2 \rangle = \frac{6}{9}\left[\lambda_1^2 + \lambda_2^2 - \lambda_1\lambda_2 + 3\lambda_1\right]. \tag{7.29}$$

The coefficient $\frac{6}{9}$ is usually omitted, since Casimir operators are defined up to a multiplicative constant.

Example 4. Eigenvalue of the quadratic Casimir operator of su(5) in the representation $[\lambda_1, \lambda_2, \lambda_3, \lambda_4]$

From

$$S_2 = \sum_{i=1}^{5} (\ell_i^2 - \rho_i^2)$$

$$\rho_1 = 4, \ \rho_2 = 3, \ \rho_3 = 2, \ \rho_4 = 1, \ \rho_5 = 0$$

$$\ell_1 = m_1 + 4, \ \ell_2 = m_2 + 3, \ \ell_3 = m_3 + 2, \ \ell_4 = m_4 + 1, \ \ell_5 = m_5$$

$$m_1 = \lambda_1 - \frac{\lambda}{5}, m_2 = \lambda_2 - \frac{\lambda}{5}, m_3 = \lambda_3 - \frac{\lambda}{5}, m_4 = \lambda_4 - \frac{\lambda}{5}, m_5 = -\frac{\lambda}{5} \tag{7.30}$$

one obtains

$$\langle C_2 \rangle = \lambda_1^2 + \lambda_2^2 + \lambda_3^2 + \lambda_4^2 - \frac{\lambda^2}{5} + 8\lambda_1 + 6\lambda_2 + 4\lambda_3 + 2\lambda_4 - 4\lambda, \tag{7.31}$$

with

$$\lambda = \lambda_1 + \lambda_2 + \lambda_3 + \lambda_4. \tag{7.32}$$

Example 5. Eigenvalue of the linear and quadratic Casimir operators of u(5) in the representation $[\lambda_1, \lambda_2, \lambda_3, \lambda_4, \lambda_5]$

Straightforward application of the formulas above give

$$\langle C_1 \rangle = S_1 = \lambda \tag{7.33}$$

$$\langle C_2 \rangle = S_2 - 4S_1$$
$$= \lambda_1^2 + \lambda_2^2 + \lambda_3^2 + \lambda_4^2 + \lambda_5^2 + 4\lambda_1 + 2\lambda_2 - 2\lambda_4. \tag{7.34}$$

For example, for the symmetric representation $[N, 0, 0, 0, 0]$,

$$\langle C_2 \rangle = N(N + 4). \tag{7.35}$$

7.4.2   The Orthogonal Algebra so(2n + 1)

To find the eigenvalues, construct the quantities

$$S_k = \sum_{i=-n}^{+n} (\ell_i^k - \rho_i^k); \qquad \rho_i = \ell_i - f_i$$

$$\begin{cases} \ell_i = f_i + n + i - \vartheta_{0i} \\ \ell_{-i} = -\ell_i + 2n - 1 \quad (i \neq 0) \\ \ell_0 = n \end{cases} \qquad \vartheta_{ji} = \begin{cases} 1 & j < i \\ 0 & j \geq i \end{cases}$$

$$f_{-i} = -f_i , \quad f_0 = 0$$
$$S_0 = S_1 = 0. \tag{7.36}$$

Construct the function

$$\varphi(z) = \sum_{k=3}^{\infty} a_k z^k, \qquad a_k = \sum_{j=2}^{k-1} \frac{(k-1)!}{j!(k-j)!} S_j. \tag{7.37}$$

Define the quantities B_p by

$$\exp(-\varphi(z)) = 1 - \sum_{p=2}^{\infty} B_p z^{p+1}, \qquad B_0 = B_1 = 0. \tag{7.38}$$

Then, the expectation value of C_p in the representation $[f_n, f_{n-1}, \ldots, f_1]$ (note the inverse order of the labels) is

$$\langle C_p \rangle = (2n+1)\, \delta_{p0} + B_p - (n+\tfrac{1}{2})\, B_{p-1} - \sum_{q=1}^{p-1} [B_q - (n+\tfrac{1}{2})\, B_{q-1}]\, n^{p-q}. \qquad (7.39)$$

Finally, convert $[f_n, f_{n-1}, \ldots, f_1]$ to the standard notation $[\mu_1, \mu_2, \ldots, \mu_n]$. This algorithm gives

$$\langle C_0 \rangle = 2n + 1$$
$$\langle C_1 \rangle = 0$$
$$\langle C_2 \rangle = S_2$$

$$\cdots \qquad (7.40)$$

Example 6. Eigenvalue of the quadratic Casimir operator of so(3) in the representation $[\mu_1]$

From

$$\langle C_2 \rangle = S_2 = \sum_{i=-1}^{+1} [\ell_i^2 - (\ell_i - f_i)^2]$$

$$\begin{cases} \ell_1 = f_1 + 1 + 1 - 1 = f_1 + 1 \\ \ell_{-1} = -f_1 - 1 + 2 - 1 = -f_1 \\ \ell_0 = 1 \end{cases} \quad \begin{matrix} f_{-1} = -f_1 \\ f_0 = 0 \end{matrix} \qquad (7.41)$$

one obtains

$$\langle C_2 \rangle = 2f_1(f_1 + 1) \qquad (7.42)$$

Converting to the standard notation $[\mu_1]$

$$\langle C_2 \rangle = 2\mu_1(\mu_1 + 1) \qquad (7.43)$$

Note that these constructions are consistent with the isomorphisms of Chap. 6. Example 10 gives for the isomorphism $su(2) \sim so(3)$ the relation $\mu_1 = \tfrac{\lambda_1}{2}$. Inserting this relation into (7.43) gives

$$\langle C_2(su(2)) \rangle = \frac{1}{2}\lambda_1(\lambda_1 + 2) \qquad (7.44)$$

as in (7.27). The additional normalization factor of two in (7.43) is usually omitted and the label μ_1 is called J in quantum mechanics textbooks. The eigenvalues of the quadratic Casimir operator of the rotation algebra (and group) is then written as

$$\langle C_2(so(3)) \rangle = J(J + 1). \qquad (7.45)$$

Example 7. Eigenvalue of the quadratic Casimir operator of so(5) in the representation $[\mu_1, \mu_2]$

From

$$\langle C_2 \rangle = S_2 = \sum_{i=-2}^{+2} (2\ell_i \, f_i - f_i^2).$$

$$\begin{cases} \ell_2 = f_2 + 3 \\ \ell_1 = f_1 + 2 \\ \ell_0 = 2 \\ \ell_{-2} = -f_2 \\ \ell_{-1} = -f_1 + 1 \end{cases} \quad \begin{cases} f_{-2} = -f_2 \\ f_{-1} = -f_1 \\ f_0 = 0 \end{cases} \qquad (7.46)$$

one obtains

$$\langle C_2 \rangle = 2 \, f_2(f_2 + 3) + 2 \, f_1(f_1 + 1). \qquad (7.47)$$

Converting to the standard notation $[\mu_1, \mu_2]$

$$\langle C_2 \rangle = 2[\mu_1(\mu_1 + 3) + \mu_2(\mu_2 + 1)]. \qquad (7.48)$$

7.4.3 The Symplectic Algebra sp(2n)

For $sp(2n)$, construct the quantities

$$S_k = \sum_{i=-n, i\neq 0}^{+n} (\ell_i^k - \rho_i^k), \qquad \rho_i = \ell_i - f_i$$

$$\begin{cases} \ell_i = f_i + n + i \\ \ell_{-i} = -\ell_i + 2n \end{cases} \qquad (7.49)$$

$$f_{-i} = -f_i, \qquad S_0 = S_1 = 0$$

and the function $\varphi(z)$ as before. Then

$$\langle C_p \rangle = 2n \, \delta_{p0} + (B_p - n \, B_{p-1}) - \sum_{q=1}^{p-1} (B_q - n \, B_{q-1}) \, (n + \frac{1}{2})^{p-q}. \qquad (7.50)$$

Finally, convert $[f_n, f_{n-1}, ..., f_1]$ to the standard notation $[\mu_1, \mu_2, ..., \mu_\nu]$. The Casimir operators of lowest order, $p \leq 2$, are

$$\langle C_0 \rangle = 2n$$

$$\langle C_1 \rangle = 0$$

$$\langle C_2 \rangle = S_2$$

$$\cdots \tag{7.51}$$

The constant $2n$ does not count as an independent Casimir operator.

Example 8. Eigenvalue of the quadratic Casimir operator of sp(2) in the representation $[\mu_1]$

From

$$\langle C_2 \rangle = S_2 = \sum_{i=-1, i \neq 0}^{1} (f_i^2 - 2\ell_i f_i)$$

$$\ell_1 = f_1 + 2, \ell_{-1} = -f_1, f_{-1} = -f_1 \tag{7.52}$$

one obtains

$$\langle C_2 \rangle = -2f_1(f_1 + 2). \tag{7.53}$$

Converting to the standard notation $[\mu_1]$

$$\langle C_2 \rangle = -2\mu_1(\mu_1 + 2). \tag{7.54}$$

Note once more the consistency with the isomorphisms $sp(2) \sim su(2)$ apart from an overall minus sign.

7.4.4 The Orthogonal Algebra so(2n)

For these algebras, construct the quantities

$$S_k = \sum_{i=-n, i \neq 0}^{+n} (\ell_i^k - \rho_i^k), \qquad \rho_i = \ell_i - f_i$$

$$\begin{cases} \ell_i = f_i + n + i - (1 + \varepsilon_i) \\ \ell_{-i} = -\ell_i + 2n - 2 \end{cases} \qquad \varepsilon_i = \begin{cases} 1 & i > 0 \\ -1 & i < 0 \\ 0 & i = 0 \end{cases}$$

$$f_{-i} = -f_i, \qquad S_0 = S_1 = 0 \tag{7.55}$$

and the function $\varphi(z)$ as before. Then

$$\langle C_p \rangle = 2n\,\delta_{p0} + (B_p - n\,B_{p-1}) - \sum_{q=1}^{p-1} (B_q - n\,B_{q-1})\,(n - \frac{1}{2})^{p-q}. \qquad (7.56)$$

Finally, convert $[f_n, f_{n-1}, \ldots, f_1]$ to $[\mu_1, \mu_2, \ldots, \mu_n]$. The eigenvalues of Casimir operators of lowest order, $p \le 2$, are

$$\langle C_0 \rangle = 2n$$
$$\langle C_1 \rangle = 0$$
$$\langle C_2 \rangle = S_2$$

$$\ldots \qquad (7.57)$$

Example 9. Eigenvalue of the quadratic Casimir operator of so(2) in the representation $[\mu_1]$

From

$$S_2 = \sum_{i=-1, i \ne 0}^{+1} (2\,l_i\,f_i - f_i^2)$$

$$\left\{ \begin{array}{l} \ell_1 = f_1 \\ \ell_{-1} = -f_1 \end{array} \right. \qquad f_{-1} = -f_1 \qquad (7.58)$$

one obtains

$$\langle C_2 \rangle = 2\,f_1^2. \qquad (7.59)$$

Converting to standard notation $[\mu_1]$

$$\langle C_2 \rangle = 2\mu_1^2. \qquad (7.60)$$

Since so(2) is Abelian it has also a linear invariant. The quadratic invariant C_2, is the square of the linear invariant C_1. In quantum mechanics textbooks, the label $\mu_1 = M$ and the factor of two are omitted.

Example 10. Eigenvalue of the quadratic Casimir operator of so(4) in the representation $[\mu_1, \mu_2]$

From

$$S_2 = \sum_{i=-2, i \ne 0}^{+2} (2\,l_i\,f_i - f_i^2)$$

$$\left\{ \begin{array}{l} \ell_2 = f_2 + 2 \\ \ell_{-2} = -f_2 \end{array} \right. \left\{ \begin{array}{l} \ell_1 = f_1 + 1 \\ \ell_{-1} = -f_1 + 1 \end{array} \right. \left\{ \begin{array}{l} f_{-1} = -f_1 \\ f_{-2} = -f_2 \end{array} \right. \qquad (7.61)$$

one obtains

$$\langle C_2 \rangle = 2 \, f_2(f_2 + 2) + 2 f_1^2. \tag{7.62}$$

Converting to standard notation $[\mu_1, \mu_2]$

$$\langle C_2 \rangle = 2 \, [\mu_1(\mu_1 + 2) + \mu_2^2]. \tag{7.63}$$

Example 11. Eigenvalue of the quadratic Casimir operator of so(6) in the representation $[\mu_1, \mu_2, \mu_3]$

In this case one has

$$\langle C_2 \rangle = 2 \, f_3(f_3 + 4) + 2 \, f_2(f_2 + 2) + 2 f_1^2 \tag{7.64}$$

and, in standard notation, $[\mu_1, \mu_2, \mu_3]$,

$$\langle C_2 \rangle = 2 \, [\mu_1(\mu_1 + 4) + \mu_2(\mu_2 + 2) + \mu_3^2]. \tag{7.65}$$

Note that for all algebras $so(2n)$ the dependence on the last quantum number μ_v is always quadratic, so that the eigenvalue for $+\mu_v$ is the same as for $-\mu_v$.

7.5 Eigenvalues of Casimir Operators of Order One and Two

The eigenvalues of Casimir operators of order $p \le 2$ of all classical Lie algebras are summarized in Table 7.1. In this table, all labels are denoted by λ_i.

Table 7.1 Eigenvalues of Casimir operators of order $p \le 2$ of all classical Lie algebras

Algebra	Labels	Order	$\langle C_2 \rangle$
$u(n)$	$[\lambda_1, \lambda_2, \ldots, \lambda_n]$	1	$\sum_{i=1}^{n} \lambda_i = \lambda$
$u(n)$	$[\lambda_1, \lambda_2, \ldots, \lambda_n]$	2	$\sum_{i=1}^{n} \lambda_i \, (\lambda_i + n + 1 - 2i)$
$su(n)$	$[\lambda_1, \lambda_2, \ldots, \lambda_{n-1}, 0]$	2	$\sum_{i=1}^{n} (\lambda_i - \frac{\lambda}{n}) (\lambda_i - \frac{\lambda}{n} + 2n - 2i)$
$so(2n + 1)$	$[\lambda_1, \lambda_2, \ldots, \lambda_n]$	2	$\sum_{i=1}^{n} 2\lambda_i \, (\lambda_i + 2n + 1 - 2i)$
$so(2n)$	$[\lambda_1, \lambda_2, \ldots, \lambda_n]$	2	$\sum_{i=1}^{n} 2\lambda_i \, (\lambda_i + 2n - 2i)$
$sp(2n)$	$[\lambda_1, \lambda_2, \ldots, \lambda_n]$	2	$\sum_{i=1}^{n} 2\lambda_i \, (\lambda_i + 2n + 2 - 2i)$

Chapter 8
Tensor Operators

8.1 Definitions

We introduce an irreducible basis $\mathcal{B}$ and write it generically as $|\ \Lambda\lambda\ \rangle$, where Λ are the labels of g and λ those of the subalgebra g'

$$\begin{vmatrix} g \supset g' \\ \downarrow \quad \downarrow \\ \Lambda \quad \lambda \end{vmatrix}. \tag{8.1}$$

The elements of g, X_σ, when acting on $\mathcal{B}$, do not lead out of $\mathcal{B}$. Thus, when acting on the basis, we obtain a linear combination of the components

$$X_\sigma \mid \Lambda\lambda\ \rangle = \sum_{\lambda'} \langle\ \Lambda\lambda' \mid X_\sigma \mid \Lambda\lambda\ \rangle \mid \Lambda\lambda'\ \rangle. \tag{8.2}$$

Example 1. The action of the elements of so(3) on its irreducible basis

From Sect. 6.13 we have

$$J_z \mid J, M\rangle = M \mid J, M\rangle$$
$$J_\pm \mid J, M\rangle = \sqrt{J(J+1) - M(M \pm 1)}\mid J, M \pm 1\rangle. \tag{8.3}$$

Definition 1. Tensor operators with respect to g

A tensor operator T_λ^Λ, sometimes written as $T(\Lambda\lambda)$, is an operator that satisfies the commutation relations

$$[X_\sigma, T_\lambda^\Lambda] = \sum_{\lambda'} \langle\ \Lambda\lambda' \mid X_\sigma \mid \Lambda\lambda\ \rangle\ T_{\lambda'}^\Lambda \tag{8.4}$$

with the elements of g.

© Springer-Verlag Berlin Heidelberg 2015
F. Iachello, *Lie Algebras and Applications*, Lecture Notes in Physics 891,
DOI 10.1007/978-3-662-44494-8_8

Example 2. Tensor operators with respect to so(3)

Tensor operators with respect to $so(3)$, T_κ^k, k = integer, κ = integer = $-k$, $\ldots, +k$, satisfy commutation relations

$$[J_z, T_\kappa^k] = \kappa T_\kappa^k$$
$$[J_\pm, T_\kappa^k] = \sqrt{k(k+1) - \kappa(\kappa \pm 1)} T_{\kappa \pm 1}^k. \tag{8.5}$$

The label k is called the rank of the tensor (not to be confused with the rank of the algebra). For $k = 0$, the tensor is called a scalar, for $k = 1$ a vector, for $k = 2$ a quadrupole tensor, etc.. A coordinate realization is

$$T_\kappa^k = \sqrt{\frac{4\pi}{2k+1}} \, Y_\kappa^k(\vartheta, \varphi). \tag{8.6}$$

8.2 Coupling Coefficients

The elements of the basis are themselves tensors. One can then form tensor products in the sense of Chap. 6. For the basis, the tensor product is denoted by

$$| \Lambda_1 \Lambda_2; a \, \Lambda_{12} \lambda_{12} \rangle = \sum_{\lambda_1, \lambda_2} \langle \Lambda_1 \lambda_1 \Lambda_2 \lambda_2 | a \, \Lambda_{12} \lambda_{12} \rangle | \Lambda_1 \lambda_1 \rangle | \Lambda_2 \lambda_2 \rangle. \tag{8.7}$$

The coefficients in the sum are called coupling coefficients or Clebsch–Gordan coefficients. Sometimes, the tensor product contains the same representation more than once. A multiplicity label, a, is introduced in these cases. The notation in (8.7) is the ket notation. The corresponding bra notation is

$$\langle \Lambda_1 \Lambda_2; a \Lambda_{12} \lambda_{12} | = \sum_{\lambda_1, \lambda_2} \langle \Lambda_1 \lambda_1 | \langle \Lambda_2 \lambda_1 | \langle a \Lambda_{12} \lambda_{12} | \Lambda_1 \lambda_1 \Lambda_2 \lambda_2 \rangle^*. \tag{8.8}$$

The coefficients are in general complex and satisfy the orthogonality relations

$$\sum_{\lambda_1, \lambda_2} \langle a \Lambda_{12} \lambda_{12} | \Lambda_1 \lambda_1 \Lambda_2 \lambda_2 \rangle^* \langle \Lambda_1 \lambda_1 \Lambda_2 \lambda_2 | a' \Lambda'_{12} \lambda'_{12} \rangle = \delta_{aa'} \delta_{\Lambda_{12} \Lambda'_{12}} \delta_{\lambda_{12} \lambda'_{12}}$$

$$\sum_{a, \Lambda_{12}, \lambda_{12}} \langle \Lambda_1 \lambda_1 \Lambda_2 \lambda_2 | a \Lambda_{12} \lambda_{12} \rangle^* \langle a \Lambda_{12} \lambda_{12} | \Lambda_1 \lambda'_1 \Lambda_2 \lambda'_2 \rangle = \delta_{\lambda_1 \lambda'_1} \delta_{\lambda_2 \lambda'_2}. \tag{8.9}$$

Multiplying (8.7) by $\langle a \Lambda_{12} \lambda_{12} | \Lambda_1 \lambda_1 \Lambda_2 \lambda_1 \rangle^*$, summing over a, Λ_{12}, λ_{12}, and using the second orthogonality relation, one obtains

$$|\Lambda_1 \lambda_1\rangle |\Lambda_2 \lambda_2\rangle = \sum_{a,\Lambda_{12},\lambda_{12}} \langle a\Lambda_{12}\lambda_{12} \mid \Lambda_1 \lambda_1 \Lambda_2 \lambda_2\rangle^* |\Lambda_1 \Lambda_2; a\Lambda_{12}\lambda_{12}\rangle, \qquad (8.10)$$

called the inverse relation.

Example 3. Clebsch–Gordan coefficients of so(3)

It is convenient here to use the notation found in most textbooks in quantum mechanics. The basis states are labelled by

$$\left| \begin{array}{cc} so(3) \supset so(2) \\ \downarrow \quad\quad \downarrow \\ J \quad\quad M \end{array} \right\rangle. \qquad (8.11)$$

There are no multiplicity labels in this case and the tensor product is written as

$$|\, J_1 J_2;\ J_{12} M_{12}\,\rangle = \sum_{M_1,M_2} \langle\, J_1 M_1 J_2 M_2 \mid J_{12} M_{12}\,\rangle\, |\, J_1 M_1\,\rangle\, |\, J_2 M_2\,\rangle \qquad (8.12)$$

Coupling coefficients are defined up to a phase. For the coefficients of $so(3)$, the Condon–Shortley phase convention is almost always used. In this convention, the coefficient with maximum $M_{12} = J_{12}$ is taken positive and real

$$\langle\, J_1 M_1 J_2 M_2 \mid J_{12},\, M_{12} = J_{12}\,\rangle \geq 0. \qquad (8.13)$$

All coefficients are then real. Instead of the Clebsch–Gordan coefficients, another coupling coefficient is often used, called Wigner 3-j symbol, related to the Clebsch–Gordan coefficient by

$$\begin{pmatrix} J_1 & J_2 & J_3 \\ M_1 & M_2 & M_3 \end{pmatrix} = \frac{(-)^{J_1-J_2-M_3}}{\sqrt{2J_3+1}} \langle\, J_1 M_1 J_2 M_2 \mid J_3, -M_3\,\rangle. \qquad (8.14)$$

8.3 Wigner–Eckart Theorem

In the evaluation of the matrix elements of a tensor operator, it is convenient to make use of a theorem, called Wigner–Eckart theorem, that states that all matrix elements can be obtained from a single one, called reduced matrix element and denoted by a double bar $\| T^\Lambda \|$,

$$\langle\, \Lambda_1 \lambda_1 \mid T_\lambda^\Lambda \mid \Lambda_2 \lambda_2\,\rangle = \sum_a \langle\, a\, \Lambda_1 \lambda_1 \mid \Lambda\, \lambda\, \Lambda_2 \lambda_2\,\rangle^* \langle\, a\, \Lambda_1 \| T^\Lambda \| \Lambda_2\,\rangle. \qquad (8.15)$$

Here a is the multiplicity label (if any). Note that different authors have often different definitions (up to a constant) of the reduced matrix elements. The inverse relation is

$$\langle a\Lambda_1 \parallel T^\Lambda \parallel \Lambda_2 \rangle = \sum_{\lambda_1,\lambda_2} \langle \Lambda\lambda\Lambda_2\lambda_2 \mid a\Lambda_1\lambda_1 \rangle \langle \Lambda_1\lambda_1 \mid T^\Lambda_\lambda \mid \Lambda_2\lambda_2 \rangle. \qquad (8.16)$$

Example 4. Wigner–Eckart theorem for so(3)

In the notation of Example 3, the Wigner–Eckart theorem is written as

$$\langle\, J_1\, M_1 \mid\, T^k_\kappa \mid J_2\, M_2 \,\rangle = \langle\, J_2\, M_2\, k\, \kappa \mid J_1\, M_1 \,\rangle \langle\, J_1 \parallel T^k \parallel J_2 \,\rangle\; \frac{1}{\sqrt{2\, J_1 + 1}}. \qquad (8.17)$$

Note the extra factor of $(2\, J_1 + 1)^{-\frac{1}{2}}$. With this definition

$$\langle\, J_1\, M_1 \mid\, T^k_\kappa \mid J_2\, M_2 \,\rangle = (-)^{J_1 - M_1} \begin{pmatrix} J_1 & k & J_2 \\ -M_1 & \kappa & M_2 \end{pmatrix} \langle\, J_1 \parallel T^k \parallel J_2 \,\rangle. \qquad (8.18)$$

The inverse relation is

$$\langle\, J_1 \parallel T^k \parallel J_2 \,\rangle = \sqrt{2\, J_1 + 1} \sum_{M_1,M_2} \langle\, J_2\, M_2\, k\, \kappa \mid J_1\, M_1 \,\rangle$$

$$\times \langle\, J_1\, M_1 \mid T^k_\kappa \mid J_2\, M_2 \,\rangle. \qquad (8.19)$$

Matrix elements of tensor operators satisfy selection rules

$$\lambda \otimes \lambda_2 \supset \lambda_1$$
$$\Lambda \otimes \Lambda_2 \supset \Lambda_1 \qquad (8.20)$$

that is the representation λ_1 must be contained in the tensor product $\lambda \otimes \lambda_2$, and the representation Λ_1 must be contained in the tensor product $\Lambda \otimes \Lambda_2$.

Example 5. Selection rules for so(3)

The matrix elements of the tensor operator T^k_κ vanish

$$\langle J_1, M_1 \mid T^k_\kappa \mid J_1 M_2 \rangle = 0 \qquad (8.21)$$

unless

$$M_1 = M_2 + \kappa, \quad \mid J_2 + k \mid \geq J_1 \geq \mid J_2 - k \mid. \qquad (8.22)$$

By making use of the Wigner–Eckart theorem, from the knowledge of one matrix element, one can compute all others.

Example 6.   Use of the Wigner–Eckart theorem for so(3)

Consider the problem of computing $\langle\, J\, M\ |\ J_+\ |\ J\, M'\,\rangle$ from the knowledge of $\langle JM\ |\ J_z\ |\ JM'\rangle$. To solve this problem, write down the Wigner–Eckart theorem for J_z

$$\langle\, J\, M\ |\ J_z\ |\ J\, M\,\rangle = (-)^{J-M}\begin{pmatrix} J & 1 & J \\ -M & 0 & M \end{pmatrix}\langle\, J\ \|\ J^{(1)}\ \|\ J\,\rangle =$$

$$= \frac{M}{\sqrt{J(J+1)(2J+1)}}\,\langle\, J\ \|\ J^{(1)}\ \|\ J\,\rangle = M, \tag{8.23}$$

and obtain

$$\langle\, J\ \|\ J^{(1)}\ \|\ J\,\rangle = \sqrt{J(J+1)(2J+1)}. \tag{8.24}$$

In writing (8.23), use has been made of the tensorial character, $k = 1$, of the angular momentum and its z-component, $\kappa = 0$, under $so(3)$.

Next, write down the Wigner–Eckart theorem for J_+

$$\langle\, J\, M\ |\ J_+\ |\ J\, M'\,\rangle = (-)^{J-M}\begin{pmatrix} J & 1 & J \\ -M & 1 & M' \end{pmatrix}\langle\, J\ \|\ J^{(1)}\ \|\ J\,\rangle. \tag{8.25}$$

Insert now the reduced matrix element obtained previously to find

$$\langle\, J\, M\ |\ J_+\ |\ J\, M'\,\rangle = \sqrt{J(J+1) - M'(M'+1)}. \tag{8.26}$$

The matrix element has selection rules $M = M' + 1$.

8.4   Nested Algebras: Racah's Factorization Lemma

In the previous sections, starting with

$$\left|\begin{matrix} g & \supset & g' \\ \downarrow & & \downarrow \\ \Lambda & & \lambda \end{matrix}\right\rangle \tag{8.27}$$

a basis for the coupled algebras

$$\left|\begin{matrix} g_1 & \oplus & g_2 & \supset & g_{12} & \supset & g'_{12} \\ \downarrow & & \downarrow & & \downarrow & & \downarrow \\ \Lambda_1 & & \Lambda_2 & & a\,\Lambda_{12} & & \lambda_{12} \end{matrix}\right\rangle \tag{8.28}$$

has been constructed

$$| \Lambda_1 \Lambda_2; a \Lambda_{12} \lambda_{12} \rangle = \sum_{\lambda_1, \lambda_2} \langle \Lambda_1 \lambda_1 \Lambda_2 \lambda_2 | a \Lambda_{12} \lambda_{12} \rangle | \Lambda_1 \lambda_1 \rangle | \Lambda_2 \lambda_2 \rangle.$$

$$(8.29)$$

Quite often, one needs to consider a further decomposition into representations of another algebra g''

$$\begin{vmatrix} g & \supset & g' & \supset & g'' \\ \downarrow & & \downarrow & & \downarrow \\ \Lambda & & \lambda & & \mu \end{vmatrix}.$$

$$(8.30)$$

The algebras g, g', g'' are called nested. The basis for coupled nested algebras

$$\begin{vmatrix} g_1 & \oplus & g_2 & \supset & g_{12} & \supset & g'_{12} & \supset & g''_{12} \\ \downarrow & & \downarrow & & \downarrow & & \downarrow & & \downarrow \\ \Lambda_1 & & \Lambda_2 & & a\Lambda_{12} & & \lambda_{12} & & \mu_{12} \end{vmatrix},$$

$$(8.31)$$

can be written as

$$| \Lambda_1 \Lambda_2; a \Lambda_{12}; \lambda_{12}; \mu_{12} \rangle =$$

$$= \sum_{\substack{\lambda_1, \lambda_2 \\ \mu_1, \mu_2}} \langle \Lambda_1 \lambda_1 \mu_1; \Lambda_2 \lambda_2 \mu_2 | a \Lambda_{12} \lambda_{12} \mu_{12} \rangle | \Lambda_1 \lambda_1 \mu_1 \rangle | \Lambda_2 \lambda_2 \mu_2 \rangle, \quad (8.32)$$

where the coefficients in the sum are called nested coupling coefficients. Their calculation is rather complex. However, Racah showed in the 1940's that nested coefficients can be split into two pieces, one for each reduction, $g \supset g'$, and $g' \supset g''$, called Racah's factorization lemma (Racah 1949).

Lemma 1. *Factorization of coupling coefficients*

$$\langle \Lambda_1 \lambda_1 \mu_1; \Lambda_2 \lambda_2 \mu_2 | a \Lambda_{12} \lambda_{12} \mu_{12} \rangle =$$

$$\langle \Lambda_1 \lambda_1 \Lambda_2 \lambda_2 | a \Lambda_{12} \lambda_{12} \rangle \langle \lambda_1 \mu_1 \lambda_2 \mu_2 | \lambda_{12} \mu_{12} \rangle. \quad (8.33)$$

The coupling coefficients for algebras of rank $\ell > 1$ are often called isoscalar factors.

Example 7. Isoscalar factors of so(4)

Consider the nested chain

$$\begin{vmatrix} so(4) & \supset & so(3) & \supset & so(2) \\ \downarrow & & \downarrow & & \downarrow \\ (\omega_1, \omega_2) & & J & & M \end{vmatrix}$$

$$(8.34)$$

The coupled basis is

$$\left| \begin{array}{ccccc} so(4)_1 & \oplus & so(4)_2 & \supset so(4)_{12} & \supset so(3)_{12} & \supset so(2)_{12} \\ \downarrow & & \downarrow & \downarrow & \downarrow & \downarrow \\ (\sigma_1,\sigma_2) & & (\zeta_1,\zeta_2) & (\omega_1,\omega_2) & J_{12} & M_{12} \end{array} \right\rangle \qquad (8.35)$$

that is

$$\sum_{\substack{J_1,J_2 \\ M_1,M_2}} \langle (\sigma_1,\sigma_2) \, J_1 \, (\zeta_1,\zeta_2) \, J_2 \mid (\omega_1,\omega_2) \, J_{12} \rangle \langle J_1 M_1 J_2 M_2 \mid J_{12} M_{12} \rangle$$

$$\times \mid (\sigma_1,\sigma_2) \, J_1 M_1 \rangle \mid (\zeta_1,\zeta_2) \, J_2 M_2 \rangle . \qquad (8.36)$$

The first factor in (8.36) is the isoscalar factor for $so(4) \supset so(3)$, while the second is the isoscalar factor for $so(3) \supset so(2)$. The isoscalar factors are sometimes written as

$$\left(\begin{array}{cc|c} (\sigma_1,\sigma_2) \, (\zeta_1,\zeta_2) & (\omega_1,\omega_2) \\ J_1 & J_2 & J_{12} \end{array} \right) \left\langle \begin{array}{cc|c} J_1 & J_2 & J_{12} \\ M_1 & M_2 & M_{12} \end{array} \right\rangle , \qquad (8.37)$$

called Wigner notation. Note again that isoscalar coefficients are defined up to an arbitrary phase.

Example 8. Coupling coefficients of su(3)

For applications to particle physics, one needs to consider the coupling coefficients of $su(3)$ in the basis

$$\left| \begin{array}{cccc} su(3) & \supset su(2) & \oplus u(1) & \supset spin(2) \\ \downarrow & \downarrow & \downarrow & \downarrow \\ (\lambda,\mu) & I & Y & I_z \end{array} \right\rangle . \qquad (8.38)$$

The coupled basis is

$$\left| \begin{array}{cccccc} su(3)_1 & \oplus & su(3)_2 & \supset su(3)_{12} & \supset [su(2) \oplus u(1)]_{12} & \supset [spin(2)]_{12} \\ \downarrow & & \downarrow & \downarrow & \downarrow & \downarrow \\ (\lambda_1,\mu_1) & & (\lambda_2,\mu_2) & (\lambda,\mu) & I,Y & I_z \end{array} \right\rangle . \qquad (8.39)$$

Using Racah's factorization lemma this can be written as

$$\sum_{\substack{I_1,Y_1,I_{z_1} \\ I_2,Y_2,I_{z_2}}} \langle (\lambda_1,\mu_1) \, I_1, Y_1 ; (\lambda_2,\mu_2) \, I_2, Y_2 \mid (\lambda,\mu) \, I, Y \rangle \times \langle I_1 I_{z_1} ; I_2 I_{z_2} \mid I, I_z \rangle$$

$$\times \mid (\lambda_1,\mu_1) \, I_1 Y_1 I_{z_1} \rangle \mid (\lambda_2,\mu_2) \, I_2 Y_2, I_{z_2} \rangle , \qquad (8.40)$$

where the first factor is the $su(3) \supset su(2) \oplus u(1)$ coefficient and the second factor is the $su(2) \supset spin(2)$ coefficient, called isospin Clebsch–Gordan coefficient. The $su(3) \supset su(2) \oplus u(1)$ coupling coefficients have been tabulated by de Swart (1963).

8.5   Adjoint Operators

Consider the matrix elements of the Hermitian conjugate of the tensor operator T_λ^Λ

$$\langle \Lambda_1\lambda_1 \mid (T_\lambda^\Lambda)^\dagger \mid \Lambda_2\lambda_2 \rangle = \langle \Lambda_2\lambda_2 \mid T_\lambda^\Lambda \mid \Lambda_1\lambda_1 \rangle^*. \qquad (8.41)$$

If T_λ^Λ is a tensor operator, $(T_\lambda^\Lambda)^\dagger$ is not. But one can show that the following operator is a tensor operator

$$\mathrm{adj}\left(T_\lambda^\Lambda\right) = (-1)^{f(\Lambda,\lambda)}\left(T_{-\lambda}^{\Lambda*}\right)^\dagger. \qquad (8.42)$$

In this expression, written generically for tensor operators with respect to

$$\left|\begin{array}{cc} g & \supset g' \\ \downarrow & \downarrow \\ \Lambda & \lambda \end{array}\right\rangle, \qquad (8.43)$$

the labels Λ^* and $-\lambda$ denote the representations of g and g' such that

$$\Lambda \otimes \Lambda^* \supset [0]$$
$$\lambda \otimes -\lambda \supset (0), \qquad (8.44)$$

where $[0]$ denotes the identity representation of g and (0) the identity representation of g'. The phase $f(\Lambda, \lambda)$ is obtained from the commutation relations with the elements of g. For all representations of $so(2n + 1)$ and $sp(2n)$, $\Lambda^* = \Lambda$. For $su(n)$, the representations Λ^* and Λ are related by the equivalence relation 2, given in Chap. 6. Also, for $so(2n)$, $-\lambda$ is the conjugate representation $[\lambda_1, \lambda_2, \ldots, -\lambda_n]$.

The operator $\mathrm{adj}\left(T_\lambda^\Lambda\right)$ is called the adjoint. It transforms under the operations of g in the same manner as T. If $\mathrm{adj}(T) = T$ the operator is called self-adjoint. The condition for self-adjointness is

$$\mathrm{adj}\left(T_\lambda^\Lambda\right) = (-1)^{f(\Lambda,\lambda)}\left(T_{-\lambda}^{\Lambda*}\right)^\dagger = T_\lambda^\Lambda. \qquad (8.45)$$

Example 9. Adjoint operators of so(3)

The tensor operator of

$$\left|\begin{array}{cc} so(3) & \supset so(2) \\ \downarrow & \downarrow \\ k & \kappa \end{array}\right\rangle \qquad (8.46)$$

adjoint to T_κ^k, $k =$integer, $\kappa =$integer, will be defined in these lecture notes as

$$\text{adj}\left(T_\kappa^k\right) = (-1)^{k-\kappa}\left(T_{-\kappa}^k\right)^\dagger \qquad \text{adj}(T) = \tilde{T}^\dagger. \tag{8.47}$$

An alternative definition and notation is often used in quantum mechanics textbooks (Messiah 1958, p. 572). The adjoint of T_κ^k is denoted by

$$S_\kappa^k = (-1)^\kappa\left(T_{-\kappa}^k\right)^\dagger. \tag{8.48}$$

The notation of Wybourne is yet different with a dagger inside the indices (Wybourne 1974). The Definition (8.47) will be used in Chap. 9, Sect. 9.7 to define annihilation operators, $\tilde{b}_{l,m} = (-)^{l-m} b_{l,-m}$, that transform in the same way as the creation operators, $b_{l,m}^\dagger$, under $so(3)$. Note that the creation operators are not self-adjoint. With products of creation and annihilation operators one can form tensor operators that are self-adjoint, in particular the angular momentum operator, L_κ^1. The situation is similar for fermion operators which transform as representations of $spin(3) \sim su(2)$ and have half-integer quantum numbers j, m

$$\left| \begin{array}{cc} spin(3) \sim su(2) \supset spin(2) \\ \downarrow \qquad\qquad \downarrow \\ j \qquad\qquad m \end{array} \right\rangle. \tag{8.49}$$

The adjoint operator of the fermion creation operator $a_{j,m}^\dagger$ will be defined in Chap. 10 as $\tilde{a}_{j,m} = (-)^{j-m} a_{j,-m}$. While for boson operators the two Definitions (8.47) and (8.48) are both possible, for fermion operators the Definition (8.48) gives rise to complex phases.

8.6 Recoupling Coefficients

In treating physical systems composed of more than two particles, one needs to couple three or more representations of g. Consider three representations $|\Lambda_1, \lambda_1\rangle, |\Lambda_2, \lambda_2\rangle, |\Lambda_3, \lambda_3\rangle$. The coupled state can be written as

$$| (\Lambda_1\Lambda_2)a_{12}\,\Lambda_{12}, \Lambda_3; a\Lambda\lambda\rangle =$$
$$= \sum_{a', a_{23}, \Lambda_{23}} \langle \Lambda_1(\Lambda_2\Lambda_3)\,a_{23}, \Lambda_{23}; a'\Lambda \,|\, (\Lambda_1\Lambda_2)\,a_{12}\,\Lambda_{12}\,\Lambda_3; a\Lambda\rangle$$
$$\times |\,\Lambda_1(\Lambda_2\Lambda_3)\,a_{23}\,\Lambda_{23}\,\Lambda_3; a'\Lambda\lambda\rangle, \tag{8.50}$$

where the coefficient in the sum is called recoupling (or Wigner) coefficient.

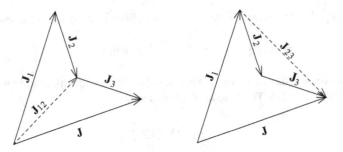

Fig. 8.1 Coupling schemes of three angular momentum vectors, $\mathbf{J}_1$, $\mathbf{J}_2$, and $\mathbf{J}_3$

Example 10. Recoupling coefficients of so(3)

For the rotation algebra $so(3)$, the recoupling coefficient is written as

$$| J_1, (J_2, J_3) J_{23} ; JM \rangle = \sum_{J_{12}} \langle (J_1 J_2) J_{12}, J_3, J \mid J_1, (J_2, J_3) J_{23}, J \rangle$$

$$\times | (J_1 J_2) J_{12}, J_3; JM \rangle. \tag{8.51}$$

It is called recoupling coefficient because it relates two possible coupling schemes, that is first couple J_1 to J_2 to give J_{12}, which is then coupled to J_3 to give the final J, or first couple J_2 and J_3 to give J_{23} which is then coupled to J_1 to give the final J. These different coupling schemes are displayed graphically in Fig. 8.1. The recoupling coefficient of $so(3)$ is usually written as

$$\langle (J_1 J_2) J_{12}, J_3, J \mid J_1, (J_2, J_3) J_{23}, J \rangle =$$

$$= (-)^{J_1 + J_2 + J_3 + J} \sqrt{(2J_{12} + 1)(2J_{23} + 1)} \begin{Bmatrix} J_1 & J_2 & J_{12} \\ J_3 & J & J_{23} \end{Bmatrix}. \tag{8.52}$$

The symbol in brackets is called a Wigner 6-j symbol.

Example 11. Recoupling coefficients of so(4)

These coefficients can be written as

$$| [(\eta_1, \eta_2) (\vartheta_1, \vartheta_2)] (\xi_1, \xi_2); (\omega_1, \omega_2); (\zeta_1, \zeta_2) \rangle =$$

$$= \sum_{\tau_1, \tau_2} \langle [(\omega_1, \omega_2) (\eta_1, \eta_2)] (\tau_1, \tau_2); (\vartheta_1, \vartheta_2); (\zeta_1, \zeta_2) |$$

$$[(\eta_1, \eta_2)(\vartheta_1, \vartheta_2)] (\xi_1, \xi_2); (\omega_1, \omega_2); (\zeta_1, \zeta_2) \rangle$$

$$\times | [(\omega_1, \omega_2) (\eta_1, \eta_2)] (\tau_1, \tau_2); (\vartheta_1, \vartheta_2); (\zeta_1, \zeta_2) \rangle. \tag{8.53}$$

8.7   Symmetry Properties of Coupling Coefficients

It is of interest to list the symmetry properties of coupling coefficients. The most important property is the symmetry under interchange of the indices 1 and 2.

$$\langle\, \Lambda_1 \lambda_1 \Lambda_2 \lambda_2 \mid a\, \Lambda\lambda \,\rangle = \varphi_1 \langle\, \Lambda_2 \lambda_2 \Lambda_1 \lambda_1 \mid a\, \Lambda\lambda \,\rangle, \tag{8.54}$$

where φ_1 is a phase. Other symmetry properties can be derived from consideration of the 3-j symbols for arbitrary Lie algebras

$$U^{\Lambda_1\Lambda_2;a_3\Lambda_3}_{\lambda_1,\lambda_2,\lambda_3} = [\Lambda_3]^{1/2} \begin{pmatrix} \Lambda_1 & \Lambda_2 & \Lambda_3 \\ \lambda_1 & \lambda_2 & \lambda_3 \end{pmatrix}_{a_3}. \tag{8.55}$$

Example 12. Symmetry properties for coupling coefficients of so(3)

The Clebsch–Gordan coefficients satisfy

$$\langle J_1 M_1 J_2 M_2 \mid JM \rangle = (-1)^{J_1 + J_2 - J} \langle J_2 M_2 J_1 M_1 \mid JM \rangle. \tag{8.56}$$

In order to display the symmetry properties of the coupling coefficients in full, it is convenient to introduce the Wigner 3-j symbols

$$\begin{pmatrix} J_1 & J_2 & J_3 \\ M_1 & M_2 & M_3 \end{pmatrix} = \frac{(-)^{J_1 - J_2 - M_3}}{\sqrt{2J_3 + 1}} \langle J_1 M_1 J_2 M_2 \mid J_3, -M_3 \rangle. \tag{8.57}$$

In terms of these symbols, the symmetry properties are

$$\begin{pmatrix} J_1 & J_2 & J_3 \\ M_1 & M_2 & M_3 \end{pmatrix} = \begin{pmatrix} J_2 & J_3 & J_1 \\ M_2 & M_3 & M_1 \end{pmatrix} = \begin{pmatrix} J_3 & J_1 & J_2 \\ M_3 & M_1 & M_2 \end{pmatrix}$$

$$= (-)^{J_1 + J_2 + J_3} \begin{pmatrix} J_1 & J_3 & J_2 \\ M_1 & M_3 & M_2 \end{pmatrix} \tag{8.58}$$

and

$$\begin{pmatrix} J_1 & J_2 & J_3 \\ -M_1 & -M_2 & -M_3 \end{pmatrix} = (-)^{J_1 + J_2 + J_3} \begin{pmatrix} J_1 & J_2 & J_3 \\ M_1 & M_2 & M_3 \end{pmatrix}. \tag{8.59}$$

8.8   How to Compute Coupling Coefficients

Explicit formulas derived by Racah are available for $so(3) \sim su(2)$. The Wigner 3-j symbol is given by the Racah formula

$$\begin{pmatrix} j_1 & j_2 & j_3 \\ m_1 & m_2 & m_3 \end{pmatrix} = \delta(m_1 + m_2 + m_3)$$

$$\times \sqrt{\frac{(j_1 + j_2 - i_3)!(j_2 + i_3 - j_1)!(j_3 + j_1 - j_2)!}{(j_1 + j_2 + j_3 + 1)!}}$$

$$\times \sqrt{(j_1 + m_1)!(j_1 - m_1)!(j_2 + m_2)!(j_2 - m_2)!}$$

$$\times \sqrt{(j_3 + m_3)!\,(j_3 - m_3)!}$$

$$\times \sum_t (-)^{j_1 - j_2 - m_3 + t} \frac{1}{t!(j_1 + j_2 - j_3 - t)!}$$

$$\times \frac{1}{(j_3 - j_2 + m_1 + t)!(j_3 - j_2 - m_2 + t)!}$$

$$\times \frac{1}{(j_1 - m_1 - t)!(j_2 + m_2 - t)!}. \tag{8.60}$$

Here $(-m)! = \infty$ when $m > 0$, $t = $ integer, $0! = 1$ and

$$t \geq 0, \qquad j_1 + j_2 - j_3 \geq t, \; -j_3 + j_2 - m_1 \leq t,$$
$$-j_3 + j_1 + m_2 \leq t, \quad j_1 - m_1 \geq t, \qquad j_2 + m_2 \geq t. \tag{8.61}$$

Since $so(4) \sim so(3) \oplus so(3) \sim su(2) \oplus su(2)$, coupling coefficients for $so(4)$ can be obtained from those of $so(3)$.

For larger algebras one has two cases: canonical and non-canonical chains. For canonical chains the construction is simple but rarely useful. For non-canonical chains it is difficult but important. The building-up principle is often used to construct coupling coefficients (isoscalar factors) for non-canonical chains. This makes use of

1. Branching rules
2. Kronecker products
3. Simple isoscalar factors (those that are zero or one)
4. Symmetry and reciprocity
5. Building-up process (start from simple and build more complex).

8.9 How to Compute Recoupling Coefficients

Explicit formulas derived by Racah are available for $so(3) \sim su(2)$. The Wigner 6-j symbol is given by

$$\begin{Bmatrix} j_1 & j_2 & j_3 \\ l_1 & l_2 & l_3 \end{Bmatrix} = \Delta (j_1 \, j_2 \, j_3) \, \Delta (j_1 \, l_2 \, l_3) \, \Delta (l_1 \, j_2 \, l_3) \, \Delta (l_1 \, l_2 \, j_3)$$

$$\times \sum_t (-)^t \, (t + 1)! \frac{1}{[t - (j_1 + j_2 + j_3)]!}$$

$$\times \frac{1}{[t - (j_1 + l_2 + l_3)]! \, [t - (l_1 + j_2 + l_3)]!}$$

$$\times \frac{1}{(j_1 + j_2 + l_1 + l_2 - t)! \, (j_2 + j_3 + l_2 + l_3 - t)!}$$

$$\times \frac{1}{(j_3 + j_1 + l_3 + l_1 - t)!}, \tag{8.62}$$

where

$$\Delta(abc) = \sqrt{\frac{(a + b - c)! (b + c - a)! (c + a - b)!}{(a + b + c + 1)!}} \tag{8.63}$$

and

$$t \ge j_1 + l_2 + j_3 \qquad , t \ge j_1 + l_2 + l_3 \qquad , t \ge l_1 + j_2 + l_3$$
$$t \ge l_1 + l_2 + j_3 \qquad , j_1 + j_2 + l_1 + l_2 \ge t \,, \, j_2 + j_3 + l_2 + l_3 \ge t . \tag{8.64}$$
$$j_3 + j_1 + l_3 + l_1 \ge t$$

For other algebras, recoupling coefficients are obtained by the building-up process.

8.10 Properties of Recoupling Coefficients

The recoupling coefficients have several interesting properties. The coefficient vanishes

$$\begin{Bmatrix} j_1 & j_2 & j_3 \\ l_1 & l_2 & l_3 \end{Bmatrix} = 0 \tag{8.65}$$

unless (j_1, j_2, l_3), (j_1, l_2, l_3), (l_1, j_2, l_3), (l_1, l_2, j_3) satisfy the triangular condition for (a, b, c), that is $| a + b | \ge c \ge | a - b |$.

They have the symmetry properties

$$\begin{Bmatrix} j_1 & j_2 & j_3 \\ l_1 & l_2 & l_3 \end{Bmatrix} = \begin{Bmatrix} j_2 & j_1 & j_3 \\ l_2 & l_1 & l_3 \end{Bmatrix} = \begin{Bmatrix} j_1 & j_3 & j_2 \\ l_1 & l_3 & l_2 \end{Bmatrix}$$

$$= \begin{Bmatrix} l_1 & l_2 & l_3 \\ j_1 & j_2 & l_3 \end{Bmatrix} = \begin{Bmatrix} j_1 & l_2 & l_3 \\ l_1 & j_2 & j_3 \end{Bmatrix}. \tag{8.66}$$

They satisfy the orthogonality relation

$$\sum_j (2j+1) \begin{Bmatrix} j_1 & j_2 & j \\ j_3 & j_4 & j' \end{Bmatrix} \begin{Bmatrix} j_1 & j_2 & j \\ j_3 & j_4 & j'' \end{Bmatrix} = \frac{\delta_{j'j''}}{2j'+1}. \tag{8.67}$$

When one of the j's is zero, they have the special value

$$\begin{Bmatrix} j_1 & j_2' & j_3 \\ j_2 & j_1' & 0 \end{Bmatrix} = \frac{(-1)^{j_1+i_2+j_3}}{\sqrt{(2j_1+1)(2j_2+1)}} \delta_{j_1 j_1'} \delta_{j_2 j_2'}. \tag{8.68}$$

The 6-j symbols are related to the 3-j symbols by

$$\begin{Bmatrix} j_1 & j_2 & j_3 \\ l_1 & l_2 & l_3 \end{Bmatrix} = \sum_{\substack{m_1 m_2 m_3 \\ m_1' m_2' m_3'}} (-1)^{j_1+l_2+j_3+l_1+l_2+l_3+m_1+m_2+m_3+m_1'+m_2'+m_3'}$$

$$\times \begin{pmatrix} j_1 & j_2 & j_3 \\ m_1 & m_2 & m_3 \end{pmatrix} \begin{pmatrix} j_1 & l_2 & l_3 \\ -m_1 & m_2' & m_3' \end{pmatrix}$$

$$\times \begin{pmatrix} l_1 & j_2 & l_3 \\ -m_1' & -m_2 & m_3' \end{pmatrix} \begin{pmatrix} l_1 & l_2 & j_3 \\ m_1' & -m_2' & -m_3 \end{pmatrix}. \tag{8.69}$$

8.11 Double Recoupling Coefficients

In treating physical systems composed of four particles, one needs double recoupling coefficients (9 $-\ j$ symbols). Introducing four representations $\mid \Lambda_1\lambda_1 \rangle$, $\mid \Lambda_2\lambda_2 \rangle, \mid \Lambda_3\lambda_3 \rangle, \mid \Lambda_4\lambda_4 \rangle$, one defines the double recoupling coefficients as

$$\mid (\Lambda_1\Lambda_3)\, a_{13}\Lambda_{13}\, (\Lambda_2\Lambda_4)\, a_{24}\Lambda_{24}\,;\, a\Lambda\lambda \rangle =$$

$$= \sum_{a', a_{12} J_{12}, a_{34} J_{34}} \langle (\Lambda_1\Lambda_2)\, a_{12}\Lambda_{12}\, (\Lambda_3\Lambda_4)\, a_{34}\Lambda_{34}\,;\, a'\Lambda\lambda \mid$$

$$\mid (\Lambda_1\Lambda_3)\, a_{13}\Lambda_{12}\, (\Lambda_2\Lambda_4)\, a_{24}\Lambda_{24}\,;\, a\Lambda\lambda \rangle$$

$$\times \mid (\Lambda_1\Lambda_2)\, a_{12}\Lambda_{12}\, (\Lambda_3\Lambda_4)\, a_{34}\Lambda_{34}\,;\, a'\Lambda\lambda \rangle. \tag{8.70}$$

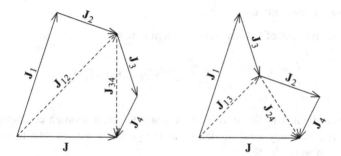

Fig. 8.2 Coupling schemes of four angular momentum vectors, $\mathbf{J}_1$, $\mathbf{J}_2$, $\mathbf{J}_3$, and $\mathbf{J}_4$

Example 13. Double recoupling coefficients of $so(3)$

The double recoupling coefficients of $so(3)$ are written as

$$| J_1 J_3 (J_{13}) J_2 J_4 (J_{24}) J M \rangle =$$

$$= \sum_{J_{12} J_{34}} \sqrt{(2J_{13}+1)(2J_{24}+1)(2J_{12}+1)(2J_{34}+1)} \begin{Bmatrix} J_1 & J_2 & J_{12} \\ J_3 & J_4 & J_{34} \\ J_{13} & J_{24} & J \end{Bmatrix}$$

$$\times | J_1 J_2 (J_{12}) J_3 J_4 (J_{34}) J M \rangle. \tag{8.71}$$

The quantity in curly bracket is called a 9-j symbol. The recoupling is shown graphically in Fig. 8.2. The 9-j symbols are related to the 6-j symbols by

$$\begin{Bmatrix} j_1 & j_2 & J \\ j_4 & j_3 & k \end{Bmatrix} = (-)^{j_2+J+j_3+k} \sqrt{(2J+1)(2k+1)} \begin{Bmatrix} j_1 & j_2 & J \\ j_3 & j_4 & J \\ k & k & 0 \end{Bmatrix}. \tag{8.72}$$

8.12 Coupled Tensor Operators

In the same way in which one couples representations, one can also couple tensors

$$\left[T^{\Lambda'} \otimes U^{\Lambda''} \right]_\lambda^{\alpha \Lambda} = \sum_{\lambda' \lambda''} \langle \Lambda' \lambda' \Lambda'' \lambda'' | \alpha \Lambda \lambda \rangle T_{\lambda'}^{\Lambda'} U_{\lambda''}^{\Lambda''}. \tag{8.73}$$

also called tensor product.

Example 14. Tensor product for so(3)

The tensor product of two tensors is written here as

$$\left[T^{k'} \times U^{k''}\right]_q^k = \sum_{q'q''} \langle k'q'k''q'' \mid k q \rangle T_{q'}^{k'} U_{q''}^{k''}, \tag{8.74}$$

where the coefficient in the sum is an ordinary Clebsch–Gordan coefficient. The symbol $\times$ is commonly used to denote tensor products with respect to $so(3)$. (A better notation would be $\otimes$.)

8.13 Reduction Formula of the First Kind

The tensor product often involves operators acting separately on systems 1 and 2.

$$T_q^k(1,2) = [T^{k_1}(1) \times T^{k_2}(2)]_q^k. \tag{8.75}$$

Using properties of the tensor product it is possible to express the matrix elements of the product in terms of matrix elements of systems 1 and 2 separately

$$\langle \alpha_1 j_1 \alpha_2 j_2 J \parallel T^k \parallel \alpha_1' j_1' \alpha_2' j_2' J' \rangle = \sqrt{(2J+1)(2k+1)(2J'+1)}$$

$$\times \begin{Bmatrix} j_1 & j_2 & J \\ j_1' & j_2' & J' \\ k_1 & k_2 & k \end{Bmatrix} \langle \alpha_1 j_1 \parallel T^{k_1} \parallel \alpha_1' j_1' \rangle \langle \alpha_2 j_2 \parallel T^{k_2} \parallel \alpha_2' j_2' \rangle. \tag{8.76}$$

Here the α's denote any additional quantum number (label). A special case is the scalar product

$$T^k \cdot U^k = (-1)^k \sqrt{2k+1} \, [T^k \times U^k]_0^0$$

$$= \sum_q (-)^q T_q^k U_{-q}^k . \tag{8.77}$$

In this case, the formula simplifies to

$$\langle \alpha_1 j_1 \alpha_2 j_2 J \parallel T^k(1) \cdot T^k(2) \parallel \alpha_1' j_1' \alpha_2' j_2' J' \rangle =$$

$$= (-1)^{j_2+J+j_1'} \sqrt{2J+1} \begin{Bmatrix} j_1 & j_2 & J \\ j_2' & j_1' & k \end{Bmatrix}$$

$$\times \langle \alpha_1 j_1 \parallel T^k \parallel \alpha_1' j_1' \rangle \langle \alpha_2 j_2 \parallel T^k \parallel \alpha_2' j_2' \rangle \delta_{JJ'}. \tag{8.78}$$

Another simple case is that in which one of the tensor operators is the identity operator. Using (8.78), one obtains

$$\langle \alpha_1 j_1 \alpha_2 j_2 J \parallel T^k(1) \parallel \alpha_1' j_1' \alpha_2' j_2' J' \rangle$$

$$= (-1)^{j_1 + J' + j_2' + k} \sqrt{(2J+1)(2J'+1)} \begin{Bmatrix} J & j_1 & j_2 \\ j_1' & J' & k \end{Bmatrix}$$

$$\langle \alpha_1 j_1 \parallel T^k(1) \parallel \alpha_1' j_1' \rangle \delta_{\alpha_2 \alpha_2'} \delta_{j_2 j_2'}. \tag{8.79}$$

8.14   Reduction Formula of the Second Kind

This is the situation when the tensor operator involves only system 1

$$T_q^k(1) = [T^{k_1}(1) \times T^{k_2}(1)]_q^k. \tag{8.80}$$

In this case one obtains

$$\langle \alpha j \parallel T^k \parallel \alpha' j' \rangle = (-)^{j+k+j'} \sqrt{(2k+1)}$$

$$\times \sum_{\alpha'' j''} \langle \alpha j \parallel T^{k_1} \parallel \alpha'' j'' \rangle \langle \alpha'' j'' \parallel T^{k_2} \parallel \alpha' j' \rangle \begin{Bmatrix} k_1 & k_2 & k \\ j' & j & j'' \end{Bmatrix}, \tag{8.81}$$

called reduction formula of the second kind.

Chapter 9
Boson Realizations

9.1 Boson Operators

Realizations of Lie algebras in terms of boson operators are of great interest for applications to a variety of problems in physics, most notably to oscillator problems in quantum mechanics and to algebraic models of rotation-vibration spectra of molecules (vibron model) and nuclei (interaction boson model).

Let $b_\alpha (\alpha = 1, .., n)$ be a set of boson operators, satisfying the commutation relations

$$[b_\alpha, b_{\alpha'}^\dagger] = \delta_{\alpha\alpha'} \qquad [b_\alpha, b_{\alpha'}] = [b_\alpha^\dagger, b_{\alpha'}^\dagger] = 0. \tag{9.1}$$

The operators b_α are often called annihilation (or destruction) operators, while their hermitian conjugates $b_\alpha^\dagger$ are called creation operators. The unitary algebra $u(n)$ can be constructed by taking bilinear products of creation and annihilation operators

$$g \doteq \quad G_{\alpha\beta} = b_\alpha^\dagger b_\beta \qquad \alpha, \beta = 1, \cdots, n. \tag{9.2}$$

The algebra is composed of n^2 elements satisfying

$$[G_{\alpha\beta}, G_{\gamma\delta}] = \delta_{\beta\gamma} G_{\alpha\delta} - \delta_{\alpha\delta} G_{\gamma\beta}. \tag{9.3}$$

(When written in this form the algebra is the real form $gl(n, R) \sim u(n)$.) A basis for the representations of $u(n)$ is

$$\mathcal{B} : \ \frac{1}{\mathcal{N}} \underbrace{b_\alpha^\dagger b_{\alpha'}^\dagger \cdots}_{N \text{ times}} | \, 0 \, \rangle \equiv | N \rangle \tag{9.4}$$

© Springer-Verlag Berlin Heidelberg 2015
F. Iachello, *Lie Algebras and Applications*, Lecture Notes in Physics 891,
DOI 10.1007/978-3-662-44494-8_9

Here $|0\rangle$ denotes a vacuum state, such that

$$b_\alpha |0\rangle = 0, \tag{9.5}$$

and $\mathcal{N}$ is a normalization. In this construction, the basis states all belong to the totally symmetric irreducible representation

$$[N] \equiv \underbrace{\square\square\ldots\square}_{N} \tag{9.6}$$

of $u(n)$ written in (9.4) as $|N\rangle$. The dimension of the basis is the dimension of the representation $[N]$, given in Chap. 6,

$$\dim[N] = \prod_{j=2,\ldots,n} \binom{N+j-1}{j-1}. \tag{9.7}$$

Thus, with boson operators it is possible to construct the Lie algebra $u(n)$, but only its symmetric representations (Bose-Einstein basis).

Any classical Lie algebra is a subalgebra of $u(n)$ (Ado's theorem) and thus can be written as a linear combination of the elements $G_{\alpha\beta}$. In the following sections an explicit construction of the Lie algebras $u(1),u(2),u(3),u(4),u(5),u(6)$ and $u(7)$ and their subalgebras will be given.

9.2 The Unitary Algebra $u(1)$

This algebra can be simply constructed with one boson operator b, satisfying

$$[b, b^\dagger] = 1 \quad [b, b] = [b^\dagger, b^\dagger] = 0. \tag{9.8}$$

The algebra is Abelian and composed of the single element

$$g \doteq b^\dagger b. \tag{9.9}$$

This operator is called the number operator, $\hat{N}$. (A hat will be placed above the operators to distinguish them from their eigenvalues.) The basis, written as

$$\left| \begin{matrix} u(1) \\ \downarrow \\ N \end{matrix} \right\rangle \tag{9.10}$$

is

$$\mathcal{B}: \quad \frac{1}{\sqrt{N!}} b^{\dagger N} \,|\,0\,\rangle \equiv |\,N\,\rangle. \tag{9.11}$$

(The capital letter N is used here, rather than the commonly used lowercase letter n, not to confuse it with the number of dimensions in $u(n)$.) In the case of $u(1)$, all representations are symmetric and thus can be constructed with boson operators. The boson operators satisfy

$$b^{\dagger}\,|N\rangle = \sqrt{N+1}\,|N+1\rangle \,; b\,|N\rangle = \sqrt{N}\,|N-1\rangle \,; b\,|0\rangle = 0. \tag{9.12}$$

Example 1. The one dimensional harmonic oscillator in quantum mechanics

The algebra $u(1)$ can be used to describe the one-dimensional harmonic oscillator. The boson operators are related to the dimensionless coordinate, x, and momentum, $p_x = -i\frac{d}{dx}$, by

$$b = \frac{1}{\sqrt{2}}\,(x + ip_x)\,, \qquad b^{\dagger} = \frac{1}{\sqrt{2}}\,(x - ip_x)\,. \tag{9.13}$$

The quantum mechanical Hamiltonian is

$$H = \frac{1}{2}\,(p_x^2 + x^2) = \frac{1}{2}\left(-\frac{d^2}{dx^2} + x^2\right). \tag{9.14}$$

When written in terms of elements of the Lie algebra $u(1)$, it reads

$$H = b^{\dagger}b + \frac{1}{2} = \hat{N} + \frac{1}{2}. \tag{9.15}$$

The representations of $u(1)$ are labelled by the integer $N = 0, 1, \ldots$. The eigenvalues of H are thus

$$E(N) = N + \frac{1}{2}, \qquad N = 0, 1, \ldots \tag{9.16}$$

The basis states are

$$|N\rangle = \frac{1}{\sqrt{N!}} b^{\dagger N}\,|0\rangle\,. \tag{9.17}$$

9.3   The Algebras $u(2)$ and $su(2)$

In this case, the index $\alpha = 1, 2$. The algebra $u(2)$ has four elements

$$g \doteq b_1^\dagger b_1, \; b_1^\dagger b_2, \; b_2^\dagger b_1, \; b_2^\dagger b_2. \tag{9.18}$$

Beginning with $u(2)$, the study of the algebraic structure of $u(n)$ requires several steps:

1. The enumeration of all possible subalgebra chains and their branchings
2. The construction of the basis $\mathcal{B}$ for all chains
3. The construction of all invariant operators and their eigenvalues

9.3.1   Subalgebra Chains

There are two possible subalgebra chains:
  Subalgebra I: $u(2) \supset u(1)$
  This is the trivial Abelian subalgebra $u(1)$

$$g' \doteq b_1^\dagger b_1, \tag{9.19}$$

leading to the canonical chain

$$\left| \begin{matrix} u(2) \supset u(1) \\ \downarrow \quad \downarrow \\ N \quad n_1 \end{matrix} \right), \tag{9.20}$$

with Gel'fand pattern

$$\left\backslash \begin{matrix} N \quad \; 0 \\ n_1 \end{matrix} \right/. \tag{9.21}$$

Here the labels of the representations are written under the algebras. Two notations are often used for canonical chains: a bra-ket notation in which only non-zero quantum numbers are displayed, and the Gel'fand notation of Chap. 6 where all quantum numbers, including zeros, are displayed. The zero in the latter notation arises from the fact that only symmetric representations can be constructed with boson operators. Using the rules of Chap. 6, one can find easily the branching

$$n_1 = 0, 1, \ldots, N. \tag{9.22}$$

The operator

$$\hat{N} = b_1^\dagger b_1 + b_2^\dagger b_2 = \hat{n}_1 + \hat{n}_2 \tag{9.23}$$

commutes with all elements of the algebra g and thus is an invariant operator of $u(2)$

$$C_1(u(2)) = \hat{N}. \tag{9.24}$$

Powers of $\hat{N}$ also commute with all elements of g and are thus also invariant operators. The quadratic Casimir operator is usually defined as

$$C_2(u(2)) = \hat{N}\left(\hat{N} + 2\right). \tag{9.25}$$

The basis is

$$\mathcal{B}: \quad \frac{1}{\sqrt{n_1! n_2!}} (b_2^\dagger)^{n_2} (b_1^\dagger)^{n_1} \mid 0 \rangle \equiv |n_1, n_2\rangle. \tag{9.26}$$

Since $n_2 = N - n_1$, this can be rewritten as

$$\frac{1}{\sqrt{n_1!(N - n_1)!}} \left(b_2^\dagger\right)^{N - n_1} \left(b_1^\dagger\right)^{n_1} |0\rangle. \tag{9.27}$$

The dimension of the irreducible representations is $N + 1$.

Example 2. The two dimensional harmonic oscillator in quantum mechanics

The algebra $u(2)$ can be used to describe the two-dimensional harmonic oscillator in (dimensionless) Cartesian coordinates, x and y. The creation and annihilation operators are related to coordinates and momenta by

$$b_1 = \frac{1}{\sqrt{2}}(x + ip_x), \quad b_2^\dagger = \frac{1}{\sqrt{2}}(x - ip_x)$$

$$b_2 = \frac{1}{\sqrt{2}}(y + ip_y), \quad b_2^\dagger = \frac{1}{\sqrt{2}}(y - ip_y). \tag{9.28}$$

The Hamiltonian of the isotropic oscillator is

$$H = \frac{1}{2}\left(p_x^2 + x^2 + p_y^2 + y^2\right). \tag{9.29}$$

It can be rewritten in terms of elements of $u(2)$ as

$$H = \hat{N} + 1 \tag{9.30}$$

with eigenvalues

$$E(N) = N + 1, \quad N = 0, 1, \ldots \tag{9.31}$$

Because of these properties, the algebra $u(2)$ is called the degeneracy algebra of the two dimensional harmonic oscillator.

   Subalgebra II: $u(2) \supset so(2)$

   It is convenient to introduce the operators

$$\overbrace{\hat{F}_+ = b_1^\dagger b_2 , \;\; \hat{F}_- = b_2^\dagger b_1 , \;\; \hat{F}_z = \frac{1}{2} (b_2^\dagger b_2 - b_1^\dagger b_1)}^{so(2)} , \;\; \hat{N} = b_2^\dagger b_2 + b_1^\dagger b_1 . \tag{9.32}$$

$$\underbrace{\phantom{\hat{F}_+ = b_1^\dagger b_2 , \;\; \hat{F}_- = b_2^\dagger b_1 , \;\; \hat{F}_z = \frac{1}{2} (b_2^\dagger b_2 - b_1^\dagger b_1)}}_{su(2)}$$

The operators $\hat{F}_+, \hat{F}_-, \hat{F}_z$ satisfy the commutation relations of $su(2)$ in the Cartan-Weyl form

$$\left[\hat{F}_+, \hat{F}_- \right] = 2\,\hat{F}_z; \quad \left[\hat{F}_z, \hat{F}_\pm \right] = \pm\,\hat{F}_\pm . \tag{9.33}$$

To make connection with the usual form, introduce

$$\hat{F}_x = \frac{1}{2} [b_1^\dagger b_2 + b_2^\dagger b_1]; \quad \hat{F}_y = \frac{1}{2i} [b_1^\dagger b_2 - b_2^\dagger b_1]. \tag{9.34}$$

The operators $\hat{F}_x, \hat{F}_y, \hat{F}_z$ satisfy the usual commutation relations

$$\left[\hat{F}_x, \hat{F}_y \right] = i\,\hat{F}_z; \quad \left[\hat{F}_y, \hat{F}_z \right] = i\,\hat{F}_x; \quad \left[\hat{F}_z, \hat{F}_x \right] = i\,\hat{F}_y . \tag{9.35}$$

The invariant Casimir operator is

$$\hat{F}^2 = \hat{F}_x^2 + \hat{F}_y^2 + \hat{F}_z^2 = \frac{1}{2}(\hat{F}_+\hat{F}_- + \hat{F}_-\hat{F}_+) + \hat{F}_z^2 = \hat{F}_-\hat{F}_+ + \hat{F}_z\,(\hat{F}_z + 1). \tag{9.36}$$

This operator commutes with all elements

$$\left[\hat{F}^2, \hat{F}_\pm \right] = \left[\hat{F}^2, \hat{F}_z \right] = 0 \tag{9.37}$$

and thus

$$\hat{F}^2 = C_2(su(2)). \tag{9.38}$$

When written in terms of boson operators

$$\hat{F}^2 = \frac{1}{4}\hat{N}(\hat{N}+2) = \frac{1}{4}(b_2^\dagger b_2 + b_1^\dagger b_1)(b_2^\dagger b_2 + b_1^\dagger b_1 + 2). \qquad (9.39)$$

The invariant $\hat{F}^2$ is related to that introduced previously in (9.25) by multiplication by $\frac{1}{4}$. This factor is due to the definition of the operators $\hat{F}$.

The basis is written in bra-ket notation as

$$\left| \begin{array}{cc} u(2) \supset so(2) \\ \downarrow \quad\quad \downarrow \\ N \quad\quad M \end{array} \right). \qquad (9.40)$$

The intermediate step $su(2)$ in the chain $u(2) \supset su(2) \supset so(2)$ may or may not be written down, since, for totally symmetric (bosonic) representations, no additional label (quantum number) is required when going from $u(n)$ to $su(n)$. Since one goes from a unitary to an orthogonal algebra, the chain $u(2) \supset so(2)$ is non-canonical. The rules of Chap. 6 give the branching

$$M = \pm N, \pm(N-2), \ldots, \pm 1 \text{ or } 0, \ N = \text{odd or even}. \qquad (9.41)$$

Note the $\pm$ sign, which arises from the fact that $so(2)$ is an orthogonal algebra in an even number of dimensions. The dimension of the representation is $N + 1$ (as in Chain I). The basis can be converted to the standard form by introducing the quantum numbers

$$F = \frac{N}{2}, \qquad F_z = \frac{M}{2}. \qquad (9.42)$$

In this form the branching is the familiar quantum mechanics result

$$F_z = -F, -F+1, \ldots, F-1, F. \qquad (9.43)$$

The dimension of the representation is $(2F + 1)$. The basis is written as

$$\left| \begin{array}{cc} su(2) \supset so(2) \\ \downarrow \quad\quad \downarrow \\ F \quad\quad F_z \end{array} \right) \qquad (9.44)$$

and can be constructed with boson operators as

$$|F, F_z\rangle = \sqrt{\frac{1}{(F+F_z)!(F-F_z)!}} (b_2^\dagger)^{F+F_z} (b_1^\dagger)^{F-F_z} |0\rangle. \qquad (9.45)$$

This construction of $su(2)$ is called the Jordan-Schwinger construction (Schwinger 1965).

Note that, since $u(1) \sim so(2)$, there is no difference between subalgebra chains I and II. They can be converted into each other by

$$\hat{F}_z = \frac{1}{2}(\hat{n}_2 - \hat{n}_1) = \frac{\hat{N}}{2} - \hat{n}_1; \qquad F_z = \frac{N}{2} - n_1$$

$$F_z = \frac{N}{2}, \frac{N}{2} - 1, \ldots, -\frac{N}{2} = F, F-1, \ldots, -F. \qquad (9.46)$$

The situation is summarized in the graph

$$
\begin{array}{c}
u(2) \\
| \\
u(1) \sim so(2)
\end{array}
\qquad (9.47)
$$

called a lattice of algebras, a concept which will become clear in the following sections.

The algebras $u(2)$ and $su(2)$ constructed with boson operators play a crucial role in applications to problems in physics: (i) $u(2)$ is the degeneracy algebra of the two dimensional harmonic oscillator; (ii) $su(2) \sim so(3)$ is the angular momentum algebra.

9.4　The Algebras $u(n), n \geq 3$

Starting with $u(3)$, the possibility arises to have the angular momentum algebra as subalgebra of $u(n)$. For applications to problems with rotational invariance, it is convenient to introduce another form of the Lie algebra $u(n)$, called the Racah form. This is obtained from the Lie algebra of (9.2) by a change of basis. The Racah form can be constructed by introducing boson operators that transform as tensor operators under $so(3)$. Since this is a generic method, it is of interest to provide a general definition.

9.4.1　Racah Form

Introduce boson creation $b^\dagger_{l,m}$ and annihilation $b_{l,m}$ operators satisfying commutation relations

$$\left[b_{l,m}, b^\dagger_{l',m'}\right] = \delta_{ll'}\delta_{mm'}$$

$$[b_{l,m}, b_{l',m'}] = \left[b^\dagger_{l,m}, b^\dagger_{l',m'}\right] = 0. \qquad (9.48)$$

If the operators $b^\dagger_{l,m}$ satisfy the commutation relations (8.5) with the elements of the algebra $so(3)$, the explicit expression of which is defined below, they transform as spherical tensors of rank l under $so(3)$ rotations. In order to construct adjoint operators that transform as spherical tensors under $so(3)$, one must use the results of Chap. 8, Example 9. The operators $b^\dagger_{l,m}$ and $\tilde{b}_{l,m}$,

$$b^\dagger_{l,m}; \quad \tilde{b}_{l,m} = (-1)^{l-m} b_{l,-m} \tag{9.49}$$

transform as spherical tensors.

The coupled bilinear products

$$G^{(k)}_\kappa (l, l') \equiv \left[b^\dagger_l \times \tilde{b}_{l'} \right]^{(k)}_\kappa$$
$$= \sum_{m,m'} \langle l m l' m' \mid k \kappa \rangle b^\dagger_{l,m} \tilde{b}_{l',m'}, \tag{9.50}$$

with $|l + l'| \geq k \geq |l - l'|$ generate the Lie algebra $u(n)$. Here $n = \sum_i (2l_i + 1)$, where l_i are the values of l used in the construction of the Lie algebra. The commutation relations of the operators $G^{(k)}_\kappa$ are

$$\left[G^{(k)}_\kappa (l, l'), G^{(k')}_{\kappa'} (l'', l''') \right] = \sum_{k'', \kappa''} (2k + 1)^{1/2} (2k' + 1)^{1/2} \langle k\kappa k'\kappa' \mid k''\kappa'' \rangle$$

$$\times \left[(-)^{l+l'''+k''} \begin{Bmatrix} l' & l''' & k' \\ k'' & k & l \end{Bmatrix} \delta_{l'l''} G^{(k'')}_{\kappa''} (l, l''') \right.$$

$$\left. - (-)^{l'+l''+k+k'} \begin{Bmatrix} l'' & l & k' \\ k & k'' & l' \end{Bmatrix} \delta_{ll'''} G^{(k'')}_{\kappa''} (l'', l') \right]. \tag{9.51}$$

This form of the algebra and of the commutation relations is known as Racah's form (Racah 1965).

9.4.2   Tensor Coupled Form of the Commutators

In deriving the preceding formula (9.51) and the corresponding formula for fermions, discussed in Chap. 10, it is useful to introduce a tensor coupled form of the commutators of the Lie algebras $u(n)$. The tensor commutator of two tensor operators $G^{(e)}$ and $G^{(f)}$ is defined by

$$\left[G^{(e)}, G^{(f)} \right]^{(g)}_\xi = \sum_{\varepsilon, \varphi} \langle e\varepsilon f\varphi \mid g\xi \rangle \left[G^{(e)}_\varepsilon, G^{(f)}_\varphi \right]. \tag{9.52}$$

This $\left[G^{(e)}, G^{(f)}\right]^{(g)}$ is itself a tensor operator and, at the tensor level, can be written as

$$\left[G^{(e)}, G^{(f)}\right]^{(g)} = (G^{(e)} \times G^{(f)})^{(g)} - (-)^{g-e-f}(G^{(f)} \times G^{(e)})^{(g)}. \tag{9.53}$$

The tensor coupled form of the commutator was introduced in French (1966) and developed in Chen (1993)

The commutator formula for the generators of $u(n)$ can be written as

$$\left[G^{(e)}(a,b), G^{(f)}(c,d)\right]^{(g)}_{\xi} = (-)^{a+d+g}(-)^{2a+2d+2g}\,\hat{e}\hat{f}\begin{Bmatrix} b & d & f \\ g & e & a \end{Bmatrix}\delta_{bc}G^{(g)}_{\xi}(a,d)$$

$$- (-)^{b+c+e+f}(-)^{2c+2d+2g}\,\hat{e}\hat{f}\begin{Bmatrix} c & a & f \\ e & g & b \end{Bmatrix}\delta_{ad}G^{(g)}_{\xi}(c,b),$$

$$\tag{9.54}$$

where $\hat{j} = (2j+1)^{1/2}$, valid for a, b, c and d *all* integer (bosonic) or *all* half-integer (fermionic). For the bosonic case $(-)^{2a+2d+2g} = (-)^{2c+2d+2g} = +1$, while for the fermionic case, to be discussed in Chap. 10, $(-)^{2a+2d+2g} = (-)^{2c+2d+2g} = -1$. The uncoupled commutators (9.51) can be obtained from (9.52) by multiplying by the Clebsch-Gordan coefficient of $so(3)$, $\langle e\varepsilon f\varphi \mid g\xi\rangle$, and summing over g and ξ

$$\left[G^{(e)}_{\varepsilon}, G^{(f)}_{\varphi}\right] = \sum_{g,\xi}\langle e\varepsilon f\varphi \mid g\xi\rangle\left[G^{(e)}, G^{(f)}\right]^{(g)}_{\xi}. \tag{9.55}$$

9.4.3 Subalgebra Chains Containing so(3)

A generic subalgebra chain (called a classification scheme) for $u(2l+1)$ constructed with boson operators can be obtained as follows:

 (i) Exclude the element with $k = 0, \kappa = 0$; this gives $su(2l+1)$.
 (ii) Retain only elements with $k =$ odd; this gives $so(2l+1)$.
 (iii) Retain the elements with $k = 1$; this gives $so(3)$.
 (iv) Retain the element with $k = 1, \kappa = 0$; this gives $so(2)$.

A generic subalgebra chain for a single value of l is

$$u(2l+1) \supset su(2l+1) \supset so(2l+1)$$

$$\supset \ldots \supset so(3) \supset so(2). \tag{9.56}$$

Dots have been inserted between $so(2l+1)$ and $so(3)$ since, for large l, there may be intermediate steps.

9.5 The Algebras $u(3)$ and $su(3)$

To construct these algebras, one needs three boson operators b_α ($\alpha = 1, 2, 3$). The 9 elements of $u(3)$ are

$$g \doteq b_\alpha^\dagger b_\beta \quad (\alpha, \beta = 1, 2, 3). \tag{9.57}$$

The algebraic analysis of $u(3)$ encounters a new feature, namely that there are here two subalgebra chains that are distinct (non-isomorphic), in contrast with the previous case in which the subalgebra chains were isomorphic. While the first of these chains, the canonical chain, can be studied as in Sect. 9.3, the second chain, which includes the angular momentum algebra $so(3)$ as a subalgebra, is best studied by introducing Racah form. In addition, several constructions of the same chain are possible. We begin with the constructions mostly used in quantum mechanical applications.

9.5.1 Subalgebra Chains

The Canonical Chain

Subalgebra I: $u(3) \supset u(2) \supset u(1)$

The canonical chain can be constructed by introducing three (singlet) boson operators b_1, b_2, b_3. The elements of the algebras in the chain are

$$
\begin{array}{ccc}
u(3) & u(2) & u(1) \\
b_1^\dagger b_1 & b_1^\dagger b_1 & b_1^\dagger b_1 \\
b_1^\dagger b_2, \; b_2^\dagger b_1 & b_1^\dagger b_2, \; b_2^\dagger b_1 & \\
b_1^\dagger b_3, \; b_3^\dagger b_1 & & \\
b_2^\dagger b_2 & b_2^\dagger b_2 & \\
b_2^\dagger b_3, \; b_3^\dagger b_2 & & \\
b_3^\dagger b_3 & &
\end{array}
\tag{9.58}
$$

The elements of $u(2)$ and $u(1)$ are obtained from $u(3)$ by deleting successively the boson operators b_3 and b_2. The basis is

$$\mathcal{B}: \quad \frac{1}{\sqrt{n_1! n_2! n_3!}} \, (b_1^\dagger)^{n_1} \, (b_2^\dagger)^{n_2} (b_3^\dagger)^{n_3} \, |0\rangle \equiv |n_1, n_2, n_3\rangle, \tag{9.59}$$

with

$$n_1 + n_2 + n_3 = N. \tag{9.60}$$

The labels can be arranged in the usual Dirac ket-notation

$$\begin{vmatrix} u(3) \supset u(2) \supset u(1) \\ \downarrow \quad\quad \downarrow \quad\quad \downarrow \\ N \quad\quad n \quad\quad n_1 \end{vmatrix},$$ (9.61)

with branching

$$n = 0,\ldots,N,$$
$$n_1 = 0,\ldots,n.$$ (9.62)

The Gel'fand pattern is

$$\left\backslash \begin{matrix} N & 0 & 0 \\ & n & 0 \\ & & n_1 \end{matrix} \right/ \equiv \left\backslash \begin{matrix} n_1 + n_2 + n_3 & & 0 & 0 \\ & n_1 + n_2 & & 0 \\ & & n_1 \end{matrix} \right/ .$$ (9.63)

The algebra has a linear invariant

$$C_1(u(3)) = \hat{N} = b_1^\dagger b_1 + b_2^\dagger b_2 + b_3^\dagger b_3,$$ (9.64)

and higher order invariants, $C_2(u(3))$ and $C_3(u(3))$, which are combinations of powers of $\hat{N}$.

Example 3. The harmonic oscillator in three dimensions in Cartesian coordinates

The algebra of $u(3)$ and its canonical chain describe the harmonic oscillator in three dimensions and in Cartesian coordinates. The dimensionless coordinates, x, y, z, and momenta, p_x, p_y, p_z, are related to the creation and annihilation operators by

$$b_1 = \frac{1}{\sqrt{2}} (x + ip_x), \quad b_1^\dagger = \frac{1}{\sqrt{2}} (x - ip_x)$$

$$b_2 = \frac{1}{\sqrt{2}} (y + ip_y), \quad b_2^\dagger = \frac{1}{\sqrt{2}} (y - ip_y)$$

$$b_3 = \frac{1}{\sqrt{2}} (z + ip_z), \quad b_3^\dagger = \frac{1}{\sqrt{2}} (z - ip_z).$$ (9.65)

The Hamiltonian of the isotropic oscillator is

$$H = \frac{1}{2} \left(p_x^2 + p_y^2 + p_z^2 + x^2 + y^2 + z^2 \right).$$ (9.66)

Its algebraic form is

$$H = \sum_{i=1}^{3} b_i^\dagger b_i + \frac{3}{2} = \hat{N} + \frac{3}{2}, \tag{9.67}$$

with eigenvalues

$$E(N) = N + \frac{3}{2}, \quad N = 0, 1, \ldots. \tag{9.68}$$

The eigenvalues depend only on the quantum number $N = n_3 + n_2 + n_1$ and not on the other quantum numbers $n = n_2 + n_1$ and n_1. The canonical chain is also useful to describe the anisotropic oscillator, with Hamiltonian

$$H = \frac{1}{2}\left(p_x^2 + p_y^2 + p_z^2 + \omega_x^2 x^2 + \omega_y^2 y^2 + \omega_z^2 z^2\right) \tag{9.69}$$

and eigenvalues

$$E(n_1, n_2, n_3) = \omega_x n_1 + \omega_y n_2 + \omega_z n_3 + \frac{3}{2},$$
$$n_1, n_2, n_3 = 0, 1, \ldots \tag{9.70}$$

The algebra $u(3)$ is the degeneracy algebra of the three dimensional harmonic oscillator.

The Non-Canonical Chain and its Racah Form

Subalgebra II: $u(3) \supset so(3) \supset so(2)$

In order to construct the non-canonical chain $u(3) \supset so(3) \supset so(2)$, it is convenient to introduce spherical boson operators (tensor operators with respect to $so(3)$)

$$p_{\pm 1}^\dagger = \mp \frac{1}{\sqrt{2}}(b_1^\dagger \pm i\, b_2^\dagger), \qquad p_0^\dagger = b_3^\dagger \tag{9.71}$$

denoted by $p_\mu^\dagger(\mu = 0, \pm 1)$ and their adjoint $\tilde{p}_\mu = (-1)^{1-\mu} p_{-\mu}$. The three boson operators $p_\mu^\dagger(\mu = \pm 1, 0)$ transform as the representation $l = 1$ of $so(3)$. In the generic construction of the elements of the algebra given in the preceding section, there is thus only one value of l.

The elements of $u(3)$ in Racah form are

$$G_0^{(0)}(pp) = (p^\dagger \times \tilde{p})_0^{(0)} \quad 1$$

$$G_\kappa^{(1)}(pp) = (p^\dagger \times \tilde{p})_\kappa^{(1)} \quad 3 \tag{9.72}$$

$$G_\kappa^{(2)}(pp) = (p^\dagger \times \tilde{p})_\kappa^{(2)} \quad 5$$

for a total of nine elements. The number of elements $(2k + 1)$ is written next to them.

Subalgebras of $u(3)$ can be then constructed as follows: (i) deleting the element $G_0^{(0)}$ gives $su(3)$. (ii) Keeping only the elements with $k = 1$ gives $so(3)$. (iii) Keeping only $G_0^{(1)}$ gives $so(2)$.

$$\begin{array}{cccc}
u(3) & su(3) & so(3) & so(2)
\end{array}$$

$$\begin{array}{l}
(p^\dagger \times \tilde{p})_0^{(0)} \\
(p^\dagger \times \tilde{p})_\kappa^{(1)} \quad (p^\dagger \times \tilde{p})_\kappa^{(1)} \quad (p^\dagger \times \tilde{p})_\kappa^{(1)} \quad (p^\dagger \times \tilde{p})_0^{(1)} \\
(p^\dagger \times \tilde{p})_0^{(2)} \quad (p^\dagger \times \tilde{p})_0^{(2)}
\end{array} \tag{9.73}$$

The basis can be written as

$$\left| \begin{array}{ccc}
u(3) \supset so(3) \supset so(2) \\
\downarrow & \downarrow & \downarrow \\
N & L & M_L
\end{array} \right\rangle . \tag{9.74}$$

Using the rules of Chap. 6, one obtains the branching

$$\begin{array}{l}
L = N, N - 2, \ldots, 1 \text{ or } 0 \quad (N = \text{ odd or even}) \\
M_L = -L, \ldots, +L
\end{array} \tag{9.75}$$

Again the intermediate step $su(3)$ in $u(3) \supset su(3) \supset so(3) \supset so(2)$ may or may not be written down, since for symmetric representations no new quantum number is needed when going from $u(3)$ to $su(3)$.

Example 4. The three dimensional harmonic oscillator in spherical coordinates

The chain $u(3) \supset so(3) \supset so(2)$ can be used to describe the harmonic oscillator in spherical coordinates, r, ϑ, φ. These coordinates are related to the Cartesian coordinates by the familiar relations $x = r \cos \varphi \sin \vartheta$, $y = r \sin \varphi \sin \vartheta$, $z = r \cos \vartheta$. The Hamiltonian is still given by (9.66), written now as

$$H = \frac{1}{2}\left(p^2 + r^2\right). \tag{9.76}$$

The algebraic form of this Hamiltonian is still

$$H = \hat{N} + \frac{3}{2},$$ (9.77)

with eigenvalues

$$E(N) = N + \frac{3}{2}, \quad N = 0, 1, \ldots$$ (9.78)

The eigenstates, given in (9.74), can be written in the form

$$|N, L, M_L\rangle = \frac{1}{\mathcal{N}} \left(p^\dagger\right)^N_{L,M_L} |0\rangle$$ (9.79)

where the product of N operators $p^\dagger$ is coupled to L and M_L, and $\mathcal{N}$ is a normalization constant.

9.5.2 Lattice of Algebras

The situation for the branchings of $u(3)$ can be summarized in the following lattice of algebras

where Cartan's notation is used on the right-hand side. The lattice must be read from top to bottom. (For bosonic systems (symmetric representations) no new quantum number is needed when going from $u(n)$ to $su(n)$ and thus the right hand side of (9.80) is written in terms of $A_2 \sim su(3)$ and $A_1 \sim su(2)$ instead of $u(3)$ and $u(2)$).

9.5.3 Boson Calculus of $u(3) \supset so(3)$

Spherical boson operators are extensively used in physics. It is therefore of interest to develop a boson calculus, for manipulations of tensor operators with respect to $so(3)$. The basic commutation relations are

$$[p_\mu, p^\dagger_{\mu'}] = \delta_{\mu\mu'} \qquad [\tilde{p}_\mu, p^\dagger_{\mu'}] = (-1)^{1-\mu} \delta_{\mu',-\mu}.$$ (9.81)

The elements of the Lie algebra $u(3)$ can be explicitly constructed using (9.50). The element $[p^\dagger \times \tilde{p}]_0^{(0)}$ is

$$
[p^\dagger \times \tilde{p}]_0^{(0)} = \sum_{\mu, \mu'} \langle 1 \; \mu \; 1 \; \mu' \mid 00 \rangle \, p_\mu^\dagger \, \tilde{p}_{\mu'} =
$$

$$
= \sum_{\mu, \mu'} (-1)^{1-\mu} \frac{1}{\sqrt{3}} \, \delta_{\mu, -\mu'} \, p_\mu^\dagger \, \tilde{p}_{\mu'} = \sum_{\mu} (-)^{1-\mu} \frac{1}{\sqrt{3}} \, p_\mu^\dagger \, \tilde{p}_{-\mu}
$$

$$
= \frac{1}{\sqrt{3}} \sum_{\mu} p_\mu^\dagger \, p_\mu = \frac{1}{\sqrt{3}} \, \hat{N}. \tag{9.82}
$$

This element is thus proportional to the number operator $\hat{N}$ that counts the number of bosons. The other elements are

$$
[p^\dagger \times \tilde{p}]_\kappa^{(k)} = \sum_{\mu, \mu'} \langle 1\mu 1\mu' \mid k\kappa \rangle p_\mu^\dagger \tilde{p}_{\mu'}. \tag{9.83}
$$

An important ingredient is often the tensor product of two creation or annihilation operators. The tensor product of two creation or annihilation operators can only have even rank, $\lambda =$ even.

Proof. Consider the product

$$
[p^\dagger \times p^\dagger]_\mu^{(\lambda)} = \sum_{\mu_1 \mu_2} \langle 1\mu_1 \, 1 \, \mu_2 \mid \lambda\mu \rangle \, p_{\mu_1}^\dagger \, p_{\mu_2}^\dagger =
$$

$$
= \sum_{\mu_1 \mu_2} \langle 1\mu_1 \, 1\mu_2 \mid \lambda\mu \rangle \, p_{\mu_2}^\dagger \, p_{\mu_1}^\dagger = \sum_{\mu_1 \mu_2} \langle 1\mu_2 \, 1\mu_1 \mid \lambda\mu \rangle \, p_{\mu_1}^\dagger \, p_{\mu_2}^\dagger
$$

$$
= \sum_{\mu_1 \mu_2} (-)^{2-\lambda} \, \langle 1\mu_1 \, 1 \, \mu_2 \mid \lambda\mu \rangle \, p_{\mu_1}^\dagger \, p_{\mu_2}^\dagger. \tag{9.84}
$$

Equating the first and last term one finds

$$
(-)^{2-\lambda} = 1; \qquad \lambda = \text{even} = 0, 2. \tag{9.85}
$$

The elements of the algebra $u(3) \supset so(3)$ are often denoted by

$$
\hat{N} = \sqrt{3}[p^\dagger \times \tilde{p}]_0^{(0)}
$$

$$
\hat{L}_\kappa = \sqrt{2}[p^\dagger \times \tilde{p}]_\kappa^{(1)}. \tag{9.86}
$$

$$
\hat{Q}_\kappa = [p^\dagger \times \tilde{p}]_\kappa^{(2)}
$$

In applications to the harmonic oscillator problem in quantum mechanics, $\hat{N}$ represents the number operator that counts the oscillator quanta. The explicit expression for the components $\hat{L}_\kappa$ is

$$\hat{L}_0 = \left(p^\dagger_{+1} \tilde{p}_{-1} - p^\dagger_{-1} \tilde{p}_{+1} \right),$$

$$\hat{L}_{-1} = -\left(p^\dagger_{-1} \tilde{p}_0 + p^\dagger_0 \tilde{p}_{-1} \right),$$

$$\hat{L}_{+1} = \left(p^\dagger_0 \tilde{p}_{+1} + p^\dagger_{+1} \tilde{p}_0 \right). \tag{9.87}$$

With the normalization (9.86), the three components $\hat{L}_\kappa$ of the operator $\hat{L}$ satisfy the commutation relations of angular momentum

$$[\hat{L}_0, \hat{L}_{\pm 1}] = \pm \hat{L}_{\pm 1} \quad , \quad [\hat{L}_{-1}, \hat{L}_{+1}] = \hat{L}_0. \tag{9.88}$$

Proof. The commutator $\left[\hat{L}_{-1}, \hat{L}_{+1} \right]$ is

$$2[(p^\dagger \times \tilde{p})]^{(1)}_{-1}, (p^\dagger \times \tilde{p})^{(1)}_{+1}] = -[p^\dagger_{-1} \tilde{p}_0 + p^\dagger_0 \tilde{p}_{-1}, \; p^\dagger_0 \tilde{p}_{+1} + p^\dagger_{+1} \tilde{p}_0] =$$

$$= -(p^\dagger_{-1} \tilde{p}_{+1} - p^\dagger_{+1} \tilde{p}_{-1})$$

$$= \sqrt{2}(p^\dagger \times \tilde{p})^{(1)}_0 = \hat{L}_0. \tag{9.89}$$

The last operator, $\hat{Q}$, is a tensor operator of rank 2 (quadrupole tensor).

9.5.4 Matrix Elements of Operators in $u(3) \supset so(3)$

Boson calculus is used to evaluate matrix elements of operators. The basis is written as

$$\mathcal{B}: \qquad |N, L, M_L\rangle = \frac{1}{\mathcal{N}} (p^\dagger)^N_{L, M_L} |0\rangle, \tag{9.90}$$

where the N boson operators have been coupled to L, M_L. There are no missing labels here. An explicit expression in terms of solid harmonics in $p^\dagger$ is also available (van Roosmalen 1982).

Reduced matrix elements of the boson operators in this basis are

$$\langle N+1, L-1 \| p^\dagger \| N, L \rangle = [(N - L + 2) L]^{1/2}$$

$$\langle N+1, L+1 \| p^\dagger \| N, L \rangle = [(N + L + 3)(L + 1)]^{1/2}. \tag{9.91}$$

The reduced matrix elements of the annihilation operators can be obtained by using the relation

$$\langle \alpha', L' \parallel \tilde{b}_l \parallel \alpha, L \rangle = (-1)^{L-L'+l} \langle \alpha, L \parallel b_l^\dagger \parallel \alpha', L' \rangle, \tag{9.92}$$

where α denotes additional labels. The matrix elements of the elements of the algebra are

$$\left\langle N, L, M_L \mid \hat{N} \mid N, L, M_L \right\rangle = N \tag{9.93}$$

and

$$\left\langle N, L \parallel \hat{L} \parallel N, L \right\rangle = [L\,(L+1)\,(2L+1)]^{1/2}$$

$$\left\langle N, L \parallel \hat{Q} \parallel N, L \right\rangle = (2N+3) \left[\frac{L(L+1)\,(2L+1)}{6(2L-1)(2L+3)} \right]^{1/2}$$

$$\left\langle N, L+2 \parallel \hat{Q} \parallel N, L \right\rangle = \left[\frac{(N-L)\,(N+L+3)(L+1)(L+2)}{(2L+3)} \right]^{1/2}. \tag{9.94}$$

Matrix elements of polynomials in the elements of the algebra can be obtained by using the reduction formulas of Chap. 8.

9.5.5 Tensor Calculus of $u(3) \supset so(3)$

An alternative way to calculate matrix elements of operators is making use of tensor calculus. The three boson creation operators $p_{\pm 1}^\dagger$, p_0 transform under $u(3)$ as the three-dimensional representation $[1, 0, 0]$ (or $(1, 0)$ of $su(3)$). They also transform as the representation $L = 1$ under $so(3)$, with component $\kappa = \pm 1, 0$. A commonly used notation for these tensors is

$$T_{1,\kappa}^{[1,0,0]} \equiv \square, \tag{9.95}$$

that is the creation operator transform as the fundamental representation of $u(3)$, often denoted as a *particle*. The boson annihilation operators $\tilde{p}_{\pm 1}$, $\tilde{p}_0$ transform instead as the three-dimensional conjugate representation $[1, 1, 0]$ (or $(1, 1)$ of $su(3)$),

$$T_{1,\kappa}^{[1,1,0]} \equiv \begin{matrix} \square \\ \square \end{matrix}. \tag{9.96}$$

This is often denoted as an *antiparticle* (or *hole*).

The elements of the algebra, that is the bilinear products $(p^\dagger \times \tilde{p})_\kappa^{(k)}$, can be simply obtained by taking tensor products, as in Chap. 6,

$$\square \otimes \begin{matrix} \square \\ \square \end{matrix} = \begin{matrix} \square \square \\ \square \end{matrix} \oplus \begin{matrix} \square \\ \square \\ \square \end{matrix} . \tag{9.97}$$

The representations on the right hand side contain

$$\begin{array}{ccc} [2, 1, 0] & \oplus & [1, 1, 1] \\ \downarrow & & \downarrow \\ (p^\dagger \times \tilde{p})_\kappa^{(1)} & (p^\dagger \times \tilde{p})_\kappa^{(0)} . \end{array} \tag{9.98}$$

$$(p^\dagger \times \tilde{p})_\kappa^{(2)}$$

The dimensions are: $\dim[2, 1, 0] = 8, \dim[1, 1, 1] = 1$ for a total of 9 elements. Note that the elements of the algebra do not transform as its fundamental representation. Under reduction of $u(3)$ to $su(3)$, the representation $[2, 1, 0]$ becomes $(2, 1)$ with $\dim(2, 1) = 8$, while the representation $[1, 1, 1]$ becomes $(0, 0)$ with $\dim(0, 0) = 1$.

The calculation of matrix elements of a generic tensor T

$$\left\langle [n', 0, 0], L', M'_L \mid T_{k, \kappa}^{[n_1, n_2, n_3]} \mid [n, 0, 0], L, M_L \right\rangle \tag{9.99}$$

is done using the techniques of Chap. 6.

The algebra $u(3)$ (and its subalgebra $su(3)$) constructed with boson operators occupy a special role in physics since $u(3)$ is the degeneracy algebra of the three-dimensional harmonic oscillator.

9.5.6 Other Boson Constructions of $u(3)$

In addition to the construction in terms of boson operators b_1, b_2, b_3 (canonical chain) and in terms of vector bosons, $p_\mu(\mu = \pm 1, 0)$, there is another boson construction of $u(3)$ of practical interest. This construction is in terms of a singlet boson, σ, and a doublet τ_x, τ_y. The doublet can be rewritten as

$$\tau_\pm = \frac{1}{\sqrt{2}} (\tau_x \pm i\tau_y)$$

$$\tau_\pm^\dagger = \frac{1}{\sqrt{2}} (\tau_x^\dagger \mp i\tau_y^\dagger) . \tag{9.100}$$

The boson operators $\tau_\pm$ are called circular boson operators. The algebra $u(3)$ is composed of nine elements:

$$\hat{n} = \left(\tau_+^\dagger \tau_+ + \tau_-^\dagger \tau_-\right)$$

$$\hat{l} = \left(\tau_+^\dagger \tau_+ - \tau_-^\dagger \tau_-\right)$$

$$\hat{Q}_+ = \sqrt{2}\left(\tau_+^\dagger \tau_-\right)$$

$$\hat{Q}_- = \sqrt{2}\left(\tau_-^\dagger \tau_+\right)$$

$$\hat{n}_s = \left(\sigma^\dagger \sigma\right)$$

$$\hat{D}_+ = \sqrt{2}\left(\tau_+^\dagger \sigma - \sigma^\dagger \tau_-\right)$$

$$\hat{D}_- = \sqrt{2}\left(-\tau_-^\dagger \sigma + \sigma^\dagger \tau_+\right)$$

$$\hat{R}_+ = \sqrt{2}\left(\tau_+^\dagger \sigma + \sigma^\dagger \tau_-\right)$$

$$\hat{R}_- = \sqrt{2}\left(\tau_-^\dagger \sigma + \sigma^\dagger \tau_+\right). \tag{9.101}$$

This construction has two possible subalgebra chains (as before).
   Subalgebra chain I: $u(3) \supset u(2) \supset so(2)$
   The subalgebra $u(2)$ is composed by the four operators

$$u(2) \doteq \hat{n}, \hat{l}, \hat{Q}_+, \hat{Q}_-. \tag{9.102}$$

The subalgebra $so(2)$ is composed of a single operator

$$so(2) \doteq \hat{l}. \tag{9.103}$$

The basis states in this chain are characterized by the quantum numbers

$$\left| \begin{array}{ccc} u(3) \supset u(2) \supset so(2) \\ \downarrow & \downarrow & \downarrow \\ N & n & l \end{array} \right\rangle, \tag{9.104}$$

with branching rules

$$n = N, N-1, \ldots, 0;$$

$$l = \pm n, \pm(n-2), \ldots, \pm 1 \text{ or } 0 \quad (n = odd \text{ or } even). \tag{9.105}$$

Subalgebra chain II: $u(3) \supset so(3) \supset so(2)$
The subalgebra $so(3)$ is composed by the three operators

$$so(3) \doteq \hat{D}_+, \hat{D}_-, \hat{l}. \tag{9.106}$$

The subalgebra $so(2)$ is composed by the single operator $\hat{l}$ as in (9.103).

The basis states in this chain are characterized by the quantum numbers

$$\left| \begin{array}{ccc} u(3) \supset so(3) \supset so(2) \\ \downarrow & \downarrow & \downarrow \\ N & \omega & l \end{array} \right\rangle, \tag{9.107}$$

with branchings

$$\omega = N, N-2, \ldots, 1 \text{ or } 0 \quad (N = odd \text{ or } even);$$
$$-\omega \leq l \leq +\omega. \tag{9.108}$$

For this subalgebra chain, a difficulty arises, since there are two $so(3)$ algebras contained in the $u(3)$ algebra of (9.101). The first is given by (9.106). The second, denoted by $\overline{so(3)}$, is composed of the operators

$$\overline{so(3)} \doteq \hat{R}_+, \hat{R}_-, \hat{l}. \tag{9.109}$$

Basis states and branchings for $\overline{so(3)}$ are similar to those of $so(3)$.

In view of the isomorphism $so(2) \sim u(1)$, the lattice of algebras for this construction is identical to that given previously in (9.80).

The construction of $u(3)$ in terms of a singlet and a doublet has practical applications in the study of vibration-rotation spectra of molecules in two-dimensions (bending vibrations) (Iachello and Oss 1996).

9.6 The Unitary Algebra $u(4)$

This algebra can be constructed by considering four boson operators, b_α, $\alpha = 1, 2, 3, 4$. The 16 elements of the algebra are

$$G_{\alpha\beta} \doteq b_\alpha^\dagger b_\beta. \tag{9.110}$$

9.6.1 Subalgebra Chains not Containing $so(3)$

The Canonical Chain

The canonical chain $u(4) \supset u(3) \supset u(2) \supset u(1)$ can be constructed trivially as before in terms of four boson operators b_1, b_2, b_3, b_4. No more will be said about this chain.

The Doublet Chain

Another chain that can be constructed simply is that in which the boson operators are divided into two doublets, b_1, b_2 and b_3, b_4. The bilinear products of creation and annihilation operators of each doublet generate a $u(2)$ algebra (Jordan-Schwinger construction), Sect. 9.2. Denoting by $u_1(2)$ and $u_2(2)$ the two $u(2)$ algebras, one has the chain $u(4) \supset u_1(2) \oplus u_2(2)$. From there on, one can use the results of Sect. 9.2. The complete subalgebra chain is $u(4) \supset u_1(2) \oplus u_2(2) \supset so_1(2) \oplus so_2(2)$.

The situation can be summarized in the lattice of algebras

$$
\begin{array}{ccc}
 & u(4) & \\
 \diagup & & \diagdown \\
u(3) & & u_1(2) \oplus u_2(2) \\
| & & | \\
u(2) & & so_1(2) \oplus so_2(2) \\
| & & \\
u(1) & &
\end{array}
\qquad (9.111)
$$

9.6.2 Subalgebra Chains Containing so(3)

We consider here explicitly non-canonical chains that contain the angular momentum algebra $so(3)$ as a subalgebra. These are particularly important in problems with rotational invariance and will be discussed here in detail. These chains can be constructed by introducing scalar, $s^\dagger$, and vector, $p_\mu^\dagger$ ($\mu = 0, \pm 1$), boson operators that transform as $l = 0$ and $l = 1$ under $so(3)$,

$$
\begin{array}{ll}
l = 0 : & s^\dagger \\
l = 1 : & p_\mu^\dagger \, (\mu = 0, \pm 1)
\end{array}
\qquad (9.112)
$$

(Although not necessary for the construction of the algebra, in applications in molecular physics the parity of these operators is chosen to be $\mathcal{P} = (-1)^l$.) The elements of $u(4)$ are, in Racah form,

$$
\begin{array}{lll}
G_0^{(0)}(ss) & = (s^\dagger \times \tilde{s})_0^{(0)} & 1 \\[4pt]
G_0^{(0)}(pp) & = (p^\dagger \times \tilde{p})_0^{(0)} & 1 \\[4pt]
G_\kappa^{(1)}(pp) & = (p^\dagger \times \tilde{p})_\kappa^{(1)} & 3 \\[4pt]
G_\kappa^{(2)}(pp) & = (p^\dagger \times \tilde{p})_\kappa^{(2)} & 5 \\[4pt]
G_\kappa^{(1)}(ps) & = (p^\dagger \times \tilde{s})_\kappa^{(1)} & 3 \\[4pt]
G_\kappa^{(1)}(sp) & = (s^\dagger \times \tilde{p})_\kappa^{(1)} & 3
\end{array}
\qquad (9.113)
$$

for a total of 16 elements. To this algebra the standard procedure of (i) constructing all possible subalgebra chains; (ii) constructing the Casimir invariants and their eigenvalues; (iii) constructing the basis $\mathcal{B}$; is then applied.

Subalgebras

There are two subalgebra chains that contain $so(3)$:

Subalgebra I: $u(4) \supset u(3) \supset so(3) \supset so(2)$

The elements of the subalgebras and their numbers are

$$
\begin{aligned}
u(3): \\
&(p^\dagger \times \tilde{p})_0^{(0)} \quad 1 \\
&(p^\dagger \times \tilde{p})_\kappa^{(1)} \quad 3 \\
&(p^\dagger \times \tilde{p})_\kappa^{(2)} \quad 5 \\
so(3): \\
&(p^\dagger \times \tilde{p})_\kappa^{(1)} \quad 3 \\
so(2): \\
&(p^\dagger \times \tilde{p})_0^{(1)} \quad 1
\end{aligned}
\tag{9.114}
$$

Subalgebra II : $u(4) \supset so(4) \supset so(3) \supset so(2)$

The elements of the subalgebras and their numbers are

$$
\begin{aligned}
so(4): \\
&(p^\dagger \times \tilde{p})_\kappa^{(1)} \quad\quad\quad\quad 3 \\
&(p^\dagger \times \tilde{s})_\kappa^{(1)} + (s^\dagger \times \tilde{p})_\kappa^{(1)} \quad 3 \\
so(3): \\
&(p^\dagger \times \tilde{p})_\kappa^{(1)} \quad\quad\quad\quad 3 \\
so(2): \\
&(p^\dagger \times \tilde{p})_0^{(1)} \quad\quad\quad\quad 1
\end{aligned}
\tag{9.115}
$$

For this subalgebra chain, a difficulty arises, similar to that discussed in the previous Sect. 9.5, as there are two $so(4)$ algebras that can be constructed with the bilinear products of s and p boson operators. The second $so(4)$ algebra, denoted by $\overline{so(4)}$ is composed of the following operators

$$
\begin{aligned}
\overline{so(4)}: \\
&\left(p^\dagger \times \tilde{p}\right)_\kappa^{(1)} \quad\quad\quad\quad 3 \\
&i\left[\left(p^\dagger \times \tilde{s}\right)_\kappa^{(1)} - \left(s^\dagger \times \tilde{p}\right)_\kappa^{(1)}\right] \quad 3 \\
so(3): \\
&\left(p^\dagger \times \tilde{p}\right)_\kappa^{(1)} \quad\quad\quad\quad 3 \\
so(2): \\
&\left(p^\dagger \times \tilde{p}\right)_0^{(1)} \quad\quad\quad\quad 1
\end{aligned}
\tag{9.116}
$$

These two possibilities are due to an inner automorphism of the Lie algebra $so(4)$.

Invariant Operators

In addition to the algebras, it is of interest to construct also its invariant (Casimir) operators. In Racah form, the invariant operators satisfy

$$[C, G_\kappa^{(k)}] = 0 \qquad \text{for any } k, \kappa. \tag{9.117}$$

The explicit form of some of the invariant operators is:

(a) Linear operators

Only unitary $u(n)$ algebras have linear invariants. They are, for $u(4)$,

$$C_1(u(4)) = G_0^{(0)}(ss) + \sqrt{3}\, G_0^{(0)}(pp) = \hat{n}_s + \hat{n}_p = \hat{N}. \tag{9.118}$$

and, for $u(3)$,

$$C_1(u(3)) = \sqrt{3}\, G_0^{(0)}(pp) = \hat{n}_p. \tag{9.119}$$

(b) Quadratic operators

The quadratic Casimir operators of $u(n)$ can be simply constructed as the square of the linear invariants. It has become customary to use as invariant operators

$$C_2(u(4)) = \hat{N}(\hat{N} + 3) \tag{9.120}$$

and

$$C_2(u(3)) = \hat{n}_p(\hat{n}_p + 2) \tag{9.121}$$

which also commute with all elements of their respective algebras.

The quadratic invariants of $so(n)$, can be written in terms of the elements $G^{(k)}$ as

$$C_2(so(3)) = G^{(1)}(pp) \cdot G^{(1)}(pp). \tag{9.122}$$

and

$$C_2(so(4)) = G^{(1)}(pp) \cdot G^{(1)}(pp) + (G^{(1)}(ps) + G^{(1)}(sp)) \cdot (G^{(1)}(ps) + G^{(1)}(sp)) \tag{9.123}$$

The dot product denotes scalar products with respect to $so(3)$, defined in Chap. 8, Sect. 8.13. A short-hand notation, often used, is

$$\hat{n}_s = (s^\dagger \times \tilde{s})^{(0)}_0$$

$$\hat{n}_p = \sqrt{3}(p^\dagger \times \tilde{p})^{(0)}_0$$

$$\hat{L} = \sqrt{2}(p^\dagger \times \tilde{p})^{(1)}_\kappa$$

$$\hat{Q} = (p^\dagger \times \tilde{p})^{(2)}_\kappa \qquad (9.124)$$

$$\hat{D} = (p^\dagger \times \tilde{s} + s^\dagger \times \tilde{p})^{(1)}_\kappa$$

$$\hat{D}' = i(p^\dagger \times \tilde{s} - s^\dagger \times \tilde{p})^{(1)}_\kappa$$

In addition to the number operators $\hat{n}_s$ and $\hat{n}_p$, there are the angular momentum operator $\hat{L}$, the quadrupole operator $\hat{Q}$ and two (hermitian) dipole operators $\hat{D}$ and $\hat{D}'$. (In quantum mechanical applications these operators are related to the coordinate and momentum vectors.) In the notation (9.124), the invariants take a simple and familiar form

$$C_2(so(3)) = \hat{L} \cdot \hat{L} \qquad\qquad C_2(so(4)) = \hat{L} \cdot \hat{L} + \hat{D} \cdot \hat{D}. \qquad (9.125)$$

Branchings

A crucial problem for application of Lie algebraic methods to problems in physics and chemistry is the classification (or branching) problem. This problem is solved using the techniques developed in Chap. 6.

Branching I

The branching of the totally symmetric representations of $u(4)$ for the chain $u(4) \supset u(3) \supset so(3) \supset so(2)$ is

$$u(4) \ [f_1, f_2, f_3, f_4] \ [N] = \overbrace{\Box \ldots \Box}^{N-\text{times}} \equiv [N, 0, 0, 0]$$

$$u(3) \ [f_1', f_2', f_3'] \quad [n_p] = \overbrace{\Box \ldots \Box}^{n_p-\text{times}} \equiv [n_p, 0, 0]$$
$$n_p = 0, 1, \ldots, N$$

$$\qquad (9.126)$$

$$so(3) \ (\mu_1') \qquad\qquad (L)$$
$$L = n_p, n_p - 2, \ldots, 1 \text{ or } 0 \ (N = \text{odd or even})$$

$$so(2) \ (\mu_1'') \qquad\qquad (M_L)$$
$$M_L = -L, \ldots, +L$$

The abstract characterization of the representations of the algebras $u(n), so(n)$, of Chap. 6, has been replaced here by quantum numbers with physical meaning: N is the total boson number, n_p is the number of p bosons, L is the angular momentum and M_L is its z-component. This branching is in part canonical, $u(4) \supset u(3)$ and $so(3) \supset so(2)$ and in part non-canonical, $u(3) \supset so(3)$. The basis is labelled by the quantum numbers

$$\left|\begin{array}{cccc} u(4) \supset u(3) \supset so(3) \supset so(2) \\ \downarrow \quad\quad \downarrow \quad\quad \downarrow \quad\quad \downarrow \\ N \quad\quad n_p \quad\quad L \quad\quad M_L \end{array}\right\rangle. \tag{9.127}$$

Branching II

For the chain $u(4) \supset so(4) \supset so(3) \supset so(2)$, the branching is

$$u(4) \quad [f_1, f_2, f_3, f_4] \quad [N] = \overbrace{\Box \ldots \Box}^{N-\text{times}} \equiv [N, 0, 0, 0]$$

$$so(4) \ (\mu_1, \mu_2) \quad\quad (\omega, 0)$$
$$\omega = N, N-2, \ldots, 1 \text{ or } 0 \ (N = \text{odd or even})$$

$$\hspace{10cm} . \tag{9.128}$$

$$so(3) \ (\mu_1') \quad\quad (L)$$
$$L = \omega, \omega - 1, \ldots, 1, 0$$

$$so(2) \ (\mu_1'') \quad\quad (M_L)$$
$$M_L = -L, \ldots, +L$$

The basis is labelled by the quantum numbers

$$\left|\begin{array}{cccc} u(4) \supset so(4) \supset so(3) \supset so(2) \\ \downarrow \quad\quad \downarrow \quad\quad \downarrow \quad\quad \downarrow \\ N \quad\quad \omega \quad\quad L \quad\quad M_L \end{array}\right\rangle. \tag{9.129}$$

The branching $u(4) \supset so(4)$ is non-canonical while the branching $so(4) \supset so(3) \supset so(2)$ is canonical.

Eigenvalues of Casimir Operators

The eigenvalues of the Casimir operators in the appropriate irreducible representations can be obtained using the rules of Chap. 7. For the unitary algebras $u(4)$ and $u(3)$ they are trivially given by

$$\langle [N] \mid C_1(u(4)) \mid [N] \rangle = N$$
$$\langle [N] \mid C_2(u(4)) \mid [N] \rangle = N(N+3)$$

$$\langle[n_p] \mid C_1(u(3)) \mid [n_p]\rangle = n_p$$
$$\langle[n_p] \mid C_2(u(3)) \mid [n_p]\rangle = n_p(n_p + 2). \tag{9.130}$$

For the orthogonal algebras $so(4)$ and $so(3)$ they are given by

$$\langle(\omega,0) \mid C_2(so(4)) \mid (\omega,0)\rangle = \omega(\omega + 2)$$
$$\langle(L) \mid C_2(so(3)) \mid (L)\rangle = L(L + 1). \tag{9.131}$$

These eigenvalues differ from those of Table 7.1 by a factor of 2, due to the different definition of C given above. Also, note that the eigenvalues of the Casimir operators of an algebra g depend only on the labels of g and not on those of the subalgebra chain $g \supset g' \supset g'' \supset \ldots$.

Lattice of Algebras

The non-canonical chains discussed above can be depicted into a lattice of algebras

$$\tag{9.132}$$

Cartan's notation with $A_3 \sim su(4)$, $A_2 \sim su(3)$ is used on the right hand side, for reasons discussed in the paragraph following (9.80). The portion of the branching which ends at $so(3) \sim B_1$

$$\tag{9.133}$$

for the representation $[N] = [4]$ is displayed in Fig. 9.1, route I (left) and route II (right). The algebra $u(4)$ constructed with s and p bosons is known as the vibron algebra (Iachello and Levine 1995). An account of the vibron algebra is also given in Frank and van Isacker (1994).

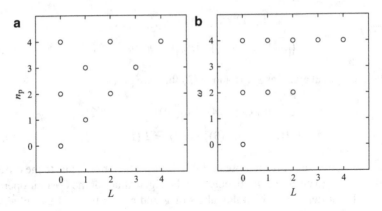

Fig. 9.1 Branchings of the representation [4] of $u(4)$. (**a**) Branching $u(4) \supset u(3) \supset so(3)$ with labels n_p and L. (**b**) Branching $u(4) \supset so(4) \supset so(3)$ with labels ω and L. The total number of states on the *left* is equal to the total number on the *right*

9.7   The Unitary Algebra $u(6)$

This algebra can be constructed by means of six boson operators, $b_\alpha, \alpha = 1, \ldots, 6$. The 36 elements of the algebra are

$$G_{\alpha\beta} \doteq b_\alpha^\dagger b_\beta \qquad\qquad \alpha, \beta = 1, \ldots, 6. \qquad\qquad (9.134)$$

9.7.1   Subalgebra Chains not Containing $so(3)$

The Canonical Chain

The canonical chain $u(6) \supset u(5) \supset u(4) \supset u(3) \supset u(2) \supset u(1)$ is trivial.

The Vector Chain

The algebra $u(6)$ can be constructed with two vector boson operators, $p_{1\mu}^\dagger$ and $p_{2\mu}^\dagger$, leading to the chain $u(6) \supset u_1(3) \oplus u_2(3)$. The algebras $u_1(3)$ and $u_2(3)$ can then be decomposed as in Sect. 9.4.

9.7.2   Subalgebra Chains Containing $so(3)$

Of particular interest for applications are the chains that can be constructed with a scalar and a quadrupole boson operator. These chains contain the angular momentum algebra $so(3)$. The boson operators $s^\dagger$ and $d_\mu^\dagger$ transform as the representation $l = 0$ and $l = 2$ of $so(3)$

$$l = 0 : \quad s^\dagger$$
$$l = 2 : \quad d_\mu^\dagger (\mu = 0, \pm 1, \pm 2). \tag{9.135}$$

The elements of the algebra $u(6)$ in Racah form are

$$
\begin{aligned}
G_0^{(0)}(ss) &= (s^\dagger \times \tilde{s})_0^{(0)} && 1 \\
G_0^{(0)}(dd) &= (d^\dagger \times \tilde{d})_0^{(0)} && 1 \\
G_\kappa^{(1)}(dd) &= (d^\dagger \times \tilde{d})_\kappa^{(1)} && 3 \\
G_\kappa^{(2)}(dd) &= (d^\dagger \times \tilde{d})_\kappa^{(2)} && 5 \\
G_\kappa^{(3)}(dd) &= (d^\dagger \times \tilde{d})_\kappa^{(3)} && 7 \\
G_\kappa^{(4)}(dd) &= (d^\dagger \times \tilde{d})_\kappa^{(4)} && 9 \\
G_\kappa^{(2)}(ds) &= (d^\dagger \times \tilde{s})_\kappa^{(2)} && 5 \\
G_\kappa^{(2)}(sd) &= (s^\dagger \times \tilde{d})_\kappa^{(2)} && 5
\end{aligned}
\tag{9.136}
$$

for a total of 36 operators. The standard procedure is then applied to this algebra.

Subalgebras

The algebra of $u(6)$ has three subalgebra chains that contain the angular momentum algebra $so(3)$.

Subalgebra I: $\quad u(6) \supset u(5) \supset so(5) \supset so(3) \supset so(2)$

The elements of the subalgebras are

$$
u(5) :
$$
$$
\begin{aligned}
(d^\dagger \times \tilde{d})_0^{(0)} && 1 \\
(d^\dagger \times \tilde{d})_\kappa^{(1)} && 3 \\
(d^\dagger \times \tilde{d})_\kappa^{(2)} && 5 \\
(d^\dagger \times \tilde{d})_\kappa^{(3)} && 7 \\
(d^\dagger \times \tilde{d})_\kappa^{(4)} && 9
\end{aligned}
$$
$$
so(5) :
$$
$$
\begin{aligned}
\left(d^\dagger \times \tilde{d}\right)_\kappa^{(1)} && 3 \\
\left(d^\dagger \times \tilde{d}\right)_\kappa^{(3)} && 7
\end{aligned}
\tag{9.137}
$$
$$
so(3) :
$$
$$
\left(d^\dagger \times \tilde{d}\right)_\kappa^{(1)} \quad 3
$$
$$
so(2) :
$$
$$
\left(d^\dagger \times \tilde{d}\right)_0^{(1)} \quad 1
$$

Subalgebra II: $u(6) \supset su(3) \supset so(3) \supset so(2)$
The elements of the subalgebras are

$u(3)$:
$$(s^\dagger \times \tilde{s})_0^{(0)} + \sqrt{5}(d^\dagger \times \tilde{d})_0^{(0)} \qquad 1$$
$$(d^\dagger \times \tilde{d})_\kappa^{(1)} \qquad 3$$
$$(d^\dagger \times \tilde{s} + s^\dagger \times \tilde{d})_\kappa^{(2)} + \tfrac{\sqrt{7}}{2}(d^\dagger \times \tilde{d})_\kappa^{(2)} \quad 5$$

$su(3)$:
$$(d^\dagger \times \tilde{d})_\kappa^{(1)} \qquad 3 .$$
$$(d^\dagger \times \tilde{s} + s^\dagger \times \tilde{d})_\kappa^{(2)} + \tfrac{\sqrt{7}}{2}(d^\dagger \times \tilde{d})_\kappa^{(2)} \quad 5 \qquad (9.138)$$

$so(3)$:
$$(d^\dagger \times \tilde{d})_\kappa^{(1)} \qquad 3$$

$so(2)$:
$$(d^\dagger \times \tilde{d})_0^{(1)} \qquad 1$$

This chain is doubled by the inner automorphism of $u(3)$, with $\overline{u(3)}$ given by

$\overline{u(3)}$:
$$\left(s^\dagger \times \tilde{s}\right)_0^{(0)} + \sqrt{5}\left(d^\dagger \times \tilde{d}\right)_0^{(0)} \qquad 1$$
$$\left(d^\dagger \times \tilde{d}\right)_\kappa^{(1)} \qquad 3 . \qquad (9.139)$$
$$\left(d^\dagger \times \tilde{s} + s^\dagger \times \tilde{d}\right)_\kappa^{(2)} - \tfrac{\sqrt{7}}{2}\left(d^\dagger \times \tilde{d}\right)_\kappa^{(2)} \quad 5$$

In applications to nuclear physics, the algebras $u(3)$ and $\overline{u(3)}$ are usually called the prolate and oblate algebras.

Subalgebra III: $u(6) \supset so(6) \supset so(5) \supset so(3) \supset so(2)$
The elements of the subalgebras are

$so(6)$:
$$(d^\dagger \times \tilde{d})_\kappa^{(1)} \qquad 3$$
$$(d^\dagger \times \tilde{d})_\kappa^{(3)} \qquad 7$$
$$(d^\dagger \times \tilde{s} + s^\dagger \times \tilde{d})_\kappa^{(2)} \quad 5$$

$so(5)$:
$$(d^\dagger \times \tilde{d})_\kappa^{(1)} \qquad 3 \qquad (9.140)$$
$$(d^\dagger \times \tilde{d})_\kappa^{(3)} \qquad 7$$

$so(3)$:
$$(d^\dagger \times \tilde{d})_\kappa^{(1)} \qquad 3$$

$so(2)$:
$$(d^\dagger \times \tilde{d})_0^{(1)} \qquad 1$$

Again this chain is doubled by the inner automorphism of $so(6)$, with $\overline{so(6)}$ given by

$$
\overline{so(6)} :
\begin{array}{ll}
\left(d^\dagger \times \tilde{d}\right)^{(1)}_\kappa & 3 \\[2mm]
\left(d^\dagger \times \tilde{d}\right)^{(3)}_\kappa & 7 \\[2mm]
i\left(d^\dagger \times \tilde{s} - s^\dagger \times \tilde{d}\right)^{(2)}_\kappa & 5
\end{array}
\tag{9.141}
$$

Invariant Operators

Invariant Casimir operators for all algebras included in the chains of the previous subsection can be constructed explicitly.

a) Linear operators

i) $u(6)$

$$
C_1(u(6)) = G_0^{(0)}(ss) + \sqrt{5}G_0^{(0)}(dd) = (s^\dagger \times s)_0^{(0)} + \sqrt{5}(d^\dagger \times \tilde{d})_0^{(0)}
$$
$$
= \hat{n}_s + \hat{n}_d \equiv \hat{N}
\tag{9.142}
$$

ii) $u(5)$

$$
C_1(u(5)) = \sqrt{5}(d^\dagger \times \tilde{d})_0^{(0)} = \hat{n}_d
\tag{9.143}
$$

iii) $u(3)$

$$
C_1(u(3)) = \hat{N}.
\tag{9.144}
$$

b) Quadratic operators

For the unitary algebras $u(6), u(5)$ and $u(3)$ the quadratic Casimir operators can be taken to be

$$
C_2(u(6)) = \hat{N}(\hat{N} + 5)
$$
$$
C_2(u(5)) = \hat{n}_d(\hat{n}_d + 4)
$$
$$
C_2(u(3)) = \hat{N}(\hat{N} + 2).
\tag{9.145}
$$

For the orthogonal algebras appearing in the branchings of $u(6)$ described in this section, they are

i) $so(3)$

$$C_2(so(3)) = G^{(1)} \cdot G^{(1)} \qquad (9.146)$$

with $G_\kappa^{(1)} \equiv G_\kappa^{(1)}(dd)$. Introducing the angular momentum operator $\hat{L}_\kappa = \sqrt{10}G_\kappa^{(1)}$, this can be rewritten as $C_2(so(3)) = \frac{1}{10}\hat{L} \cdot \hat{L}$. (The dot denotes scalar products.)

ii) $so(5)$

$$C_2(so(5)) = G^{(1)} \cdot G^{(1)} + G^{(3)} \cdot G^{(3)} \qquad (9.147)$$

with $G_\kappa^{(1)}$ as before and $G_\kappa^{(3)} \equiv G_\kappa^{(3)}(dd)$.

For the special unitary algebra $su(3)$ appearing in the branching of $u(6)$, they are

iii) $su(3)$

$$C_2(su(3)) = G^{(1)} \cdot G^{(1)} + \frac{4}{15}\tilde{G}^{(2)} \cdot \tilde{G}^{(2)}, \qquad (9.148)$$

where

$$\tilde{G}_\kappa^{(2)} = G_\kappa^{(2)}(ds) + G_\kappa^{(2)}(sd) \mp \frac{\sqrt{7}}{2}G_\kappa^{(2)}(dd), \qquad (9.149)$$

If, in addition to the angular momentum operator $\hat{L}$, one introduces the quadrupole operator $\hat{Q} = \sqrt{8}\tilde{G}^{(2)}$, the Casimir operator of $su(3)$ can be rewritten as

$$C_2(su(3)) = \frac{1}{30}\left[3\hat{L} \cdot \hat{L} + \hat{Q} \cdot \hat{Q}\right]. \qquad (9.150)$$

Branchings

Branching I

The branching of representations of $u(6)$ into representations of its subalgebra present a challenge not encountered when dealing with $u(n)$, $n \leq 4$. One needs a hidden quantum number to characterize uniquely the basis. Finding this quantum number is one of the most difficult problems in algebraic theory. The branching $u(6) \supset u(5) \supset so(5) \supset so(3) \supset so(2)$ is

$$u(6) \quad [N] \equiv \overbrace{\square\square\ldots\square}^{N-\text{times}} \equiv [N,0,0,0,0,0]$$

$$u(5) \quad [n_d] \equiv \overbrace{\square\square\ldots\square}^{n_d-\text{times}} \equiv [n_d,0,0,0,0] \quad n_d = N, N-1, \ldots, 0$$

$$so(5) \quad (v) \equiv (v, 0) \qquad\qquad\qquad v = n_d, n_d - 2, \ldots, 1 \text{ or } 0$$
$$\qquad\qquad\qquad\qquad\qquad\qquad\qquad (n_d = \text{odd or even})$$

$$so(3) \quad L \qquad\qquad\qquad\qquad\qquad\qquad \text{Algorithm 1}$$

$$so(2) \quad M_L \qquad\qquad\qquad\qquad\qquad\qquad -L \leq M_L \leq +L$$

$$\text{(9.151)}$$

The step from $so(5)$ to $so(3)$ is non-canonical and thus requires the development of an algorithm to find the values of L contained in each representation n_d.

Algorithm 1. Partition n_d as

$$n_d = 2n_\beta + 3n_\Delta + \lambda \tag{9.152}$$

where

$$n_\beta = (n_d - v)/2; \quad n_\beta = 0, 1, \ldots, \frac{n_d}{2} \text{ or } \frac{n_d - 1}{2}. \tag{9.153}$$

Then

$$L = \lambda, \lambda + 1, \lambda + 2, \ldots, 2\lambda - 2, 2\lambda. \tag{9.154}$$

[Note that $2\lambda - 1$ is missing!]. The additional quantum number $n_\Delta \doteq 0, 1, \ldots$ is a missing label (hidden quantum number) needed to characterize uniquely the decompositions of representations of $so(5)$ into representations of $so(3)$. This gives Table 9.1.

The complete classification for branching I is

$$\left| \begin{array}{ccccc} u(6) \supset u(5) \supset so(5) \supset so(3) \supset so(2) \\ \downarrow \quad\quad \downarrow \quad\quad \downarrow \quad\quad \downarrow \quad\quad \downarrow \\ N \quad\quad n_d \quad\quad v, n_\Delta \quad\quad L \quad\quad M_L \end{array} \right). \tag{9.155}$$

The total number of labels, including the missing label n_Δ, is six.

Branching II

The branching $u(6) \supset su(3) \supset so(3) \supset so(2)$ is

Table 9.1 Decomposition of totally symmetric representations of $u(6) \supset u(5)$ into representations of $so(3)$

$u(6)$ $\supset$	$u(5)$ $\supset$	$so(5)$ $\supset$		$so(3)$
N	n_d	ν	n_Δ	L
0	0	0	0	0
1	0	0	0	0
	1	1	0	2
2	0	0	0	0
	1	1	0	2
	2	2	0	4, 2
		0	0	0
3	0	0	0	0
	1	1	0	2
	2	2	0	4, 2
		0	0	0
	3	3	0	6, 4, 3
			1	0
		1	0	2

$$\overset{N\text{--times}}{\overbrace{}}$$

$$u(6)\;\;[N] \equiv \overbrace{\square\square \ldots \square}^{N\text{--times}} \equiv [N,0,0,0,0,0]$$

$$su(3)\;\;(\lambda,\mu) \qquad\qquad\qquad \text{Algorithm 2} \qquad\qquad (9.156)$$

$$so(3)\;\;L. \qquad\qquad\qquad\qquad \text{Algorithm 3}$$

$$so(2)\;\;M_L \qquad\qquad\qquad\qquad -L \le M_L \le +L$$

Here, the so-called Elliott quantum numbers, $\lambda = f_1 - f_2, \mu = f_2$, are used instead of the entries in the Young tableau (f_1, f_2) in order to conform with commonly used notation. Both steps from $u(6)$ to $su(3)$ and from $su(3)$ to $so(3)$ are non-canonical. The step from $su(3)$ to $so(3)$ is not fully reducible.

Algorithm 2. The algorithm to find the values of (λ, μ) contained in $[N]$ is

$$(2N,0) \oplus (2N-4,2) \oplus \cdots \quad \oplus \begin{cases} (0,N) & N = \text{even} \\ (2,N-1) & N = \text{odd} \end{cases} \oplus$$

$$\oplus (2N-6) \oplus (2N-10,2) \oplus \cdots \quad \oplus \begin{cases} (0,N-3) & N-3 = \text{even} \\ (2,N-4) & N-3 = \text{odd} \end{cases} \oplus$$

$$\oplus (2N-12,0) \oplus (2N-16,2) \oplus \cdots \oplus \begin{cases} (0,N-6) & N-6 = \text{even} \\ (2,N-7) & N-6 = \text{odd} \end{cases} \oplus$$

$$\oplus \cdots$$

$$(9.157)$$

Algorithm 3. The algorithm to find the values of L contained in (λ, μ) is

$$L = K, K+1, K+2, \ldots, (K + \max\{\lambda, \mu\}) \qquad (9.158)$$

Table 9.2 Decomposition of totally symmetric representations of $u(6) \supset su(3)$ into representations of $so(3)$

$u(6)$	$\supset$	$su(3)$	$\supset$	$so(3)$
N		(λ, μ)	K	L
0		$(0,0)$	0	0
1		$(2,0)$	0	$2,0$
2		$(4,0)$	0	$4,2,0$
		$(0,2)$	0	$2,0$
3		$(6,0)$	0	$6,4,2,0$
		$(2,2)$	0	$2,0$
			2	$2,3,4$
		$(0,0)$	0	0

where

$$K = \text{integer} = \min\{\lambda, \mu\}, \min\{\lambda, \mu\} - 2, \ldots, 1 \text{ or } 0 \qquad (9.159)^{*}$$

with exception of $K = 0$ for which

$$L = \max\{\lambda, \mu\}, \max\{\lambda, \mu\} - 2, \ldots, 1 \text{ or } 0. \qquad (9.160)$$

Here K is the missing label. This gives Table 9.2.

The complete classification for branching II is

$$\left|\begin{array}{cccc} u(6) \supset & su(3) & \supset so(3) \supset so(2) \\ \downarrow & \downarrow & \downarrow \qquad \downarrow \\ N & (\lambda; \mu) \ K & L \qquad M_L \end{array}\right). \qquad (9.161)$$

Again, a total of six labels is needed.

Branching III

The branching $u(6) \supset so(6) \supset so(5) \supset so(3) \supset so(2)$ is

$$u(6) \ [N] \equiv \overbrace{\Box\Box \ldots \Box}^{N-\text{times}} \equiv [N, 0, 0, 0, 0, 0]$$

$$so(6) \ \sigma \equiv (\sigma, 0, 0) \qquad\qquad\qquad \sigma = N, N - 2, \ldots, 1 \text{ or } 0$$
$$\qquad\qquad\qquad\qquad\qquad\qquad (N = \text{odd or even})$$
$$so(5) \ \tau \equiv (\tau, 0) \qquad\qquad\qquad\qquad \tau = \sigma, \sigma - 1, \ldots, 0 \qquad\qquad (9.162)$$

$$so(3) \ L \qquad\qquad\qquad\qquad\qquad\qquad \text{Algorithm 4}$$

$$so(2) \ M_L \qquad\qquad\qquad\qquad\qquad\qquad -L \le M_L \le +L$$

Table 9.3 Decomposition of totally symmetric representations of $u(6) \supset so(6)$ into representations of $so(3)$

$u(6)$	$\supset$	$so(6)$	$\supset$	$so(5)$	$\supset$	$so(3)$
N		σ		τ	ν_Δ	L
0		0		0	0	0
1		1		1	0	2
				0	0	0
2		2		2	0	4, 2
				1	0	2
				0	0	0
		0		0	0	0
3		3			0	6, 4, 3
					1	0
				2	1	4, 2
				1	0	2
				0	0	0
		1		1	0	2
				0	0	0

Algorithm 4. The algorithm to find the values of L contained in each representation τ is: Partition τ as

$$\tau = 3\nu_\Delta + \lambda, \qquad \nu_\Delta = 0, 1, \ldots \tag{9.163}$$

and take

$$L = 2\lambda, 2\lambda - 2, \ldots, \lambda + 1, \lambda. \tag{9.164}$$

[Note again that $2\lambda - 1$ is missing!]. The missing label is here ν_Δ.

This gives Table 9.3.

The complete classification for branching III is

$$\left|
\begin{array}{ccccc}
u(6) \supset & so(6) \supset & so(5) \supset & so(3) \supset & so(2) \\
\downarrow & \downarrow & \downarrow & \downarrow & \downarrow \\
N & \sigma & \tau, \nu_\Delta & L & M_L
\end{array}
\right\rangle \tag{9.165}$$

with six quantum numbers as before.

In all three chains, the total number of labels needed to characterize uniquely the totally symmetric representations of $u(6)$ is six, as one can see by considering the Gel'fand pattern

$$
\begin{matrix}
n_1 & 0 & 0 & 0 & 0 & 0 \\
 & n_2 & 0 & 0 & 0 & 0 \\
 & & n_3 & 0 & 0 & 0 \\
 & & & n_4 & 0 & 0 \\
 & & & & n_5 & 0 \\
 & & & & & n_6
\end{matrix}
\tag{9.166}
$$

Eigenvalues of Casimir Operators

The eigenvalues of Casimir operators in the representations labeled by the quantum numbers of the previous subsection are

$$
\begin{aligned}
\langle [N] \mid C_1(u(6)) \mid [N] \rangle &= N \\
\langle [N] \mid C_2(u(6)) \mid [N] \rangle &= N(N + 5) \\
\langle [n_d] \mid C_1(u(5)) \mid [n_d] \rangle &= n_d \\
\langle [n_d] \mid C_2(u(5)) \mid [n_d] \rangle &= n_d(n_d + 4) \\
\langle (\lambda, \mu) \mid C_2(su(3)) \mid (\lambda, \mu) \rangle &= \lambda^2 + \mu^2 + \lambda\mu + 3\lambda + 3\mu \\
\langle (\sigma, 0, 0) \mid C_2(so(6)) \mid (\sigma, 0, 0) \rangle &= \sigma(\sigma + 4) \\
\langle (\tau, 0) \mid C_2(so(5)) \mid (\tau, 0) \rangle &= \tau(\tau + 3) \\
\langle (L) \mid C_2(so(3)) \mid (L) \rangle &= L(L + 1).
\end{aligned}
\tag{9.167}
$$

Lattice of Algebras

The lattice of algebras is

$$
\begin{array}{ccccccc}
& u(6) & & & & A_5 & \\
\diagup & | & \diagdown & & \diagup & | & \diagdown \\
u(5) \quad so(6) & & su(3) & A_4 & D_3 & & A_2 \\
\diagdown \quad \diagup & & & \diagdown & \diagup & & \\
so(5) \quad \diagup & & & B_2 \quad \diagup & & & \\
\diagdown & & & \diagdown & & & \\
so(3) & & & B_1 & & & \\
| & & & | & & & \\
so(2) & & & D_1 & & &
\end{array}
\tag{9.168}
$$

Again, on the right hand side, Cartan notation $A_l \sim su(l + 1)$, is used [see (9.80)].

The algebra of $u(6)$ constructed with s and d bosons is known as the interacting boson model algebra (Iachello and Arima 1987). An account of the interacting boson model algebra is also given in Frank and van Isacker (1994).

9.8   The Unitary Algebra $u(7)$

In the previous sections, the unitary algebras $u(1), u(2), \ldots, u(6)$ have been constructed with bilinear products of boson operators $b_\alpha^\dagger b_\beta$. Unitary algebras $u(n)$ with $n > 6$ can be constructed in a similar way. In this section, as a last example of an explicit boson construction, the algebra $u(7)$ will be considered. This algebra is composed of the 49 elements

$$G_{\alpha\beta} \doteq b_\alpha^\dagger b_\beta \quad (\alpha, \beta = 1, 2, \ldots, 7). \tag{9.169}$$

The algebra $u(7)$ is interesting for two reasons: (i) It is the unitary algebra of lowest rank that contains as a subalgebra one of the exceptional algebras, g_2, and (ii) it has applications to the three-body problem in quantum mechanics. The subalgebra chains that describe these two situations are discussed in the following subsections.

9.8.1   Subalgebra Chain Containing g_2

This chain can be constructed by introducing an octupole boson $f^\dagger$ that transforms as the representation $l = 3$ of $so(3)$

$$l = 3: \quad f_\mu^\dagger \quad (\mu = 0, \pm 1, \pm 2, \pm 3). \tag{9.170}$$

The elements of the algebra $u(7)$ in Racah form are

$$
\begin{aligned}
G_0^{(0)}(ff) &= \left(f^\dagger \times \tilde{f} \right)_0^{(0)} \quad 1 \\
G_\kappa^{(1)}(ff) &= \left(f^\dagger \times \tilde{f} \right)_\kappa^{(1)} \quad 3 \\
G_\kappa^{(2)}(ff) &= \left(f^\dagger \times \tilde{f} \right)_\kappa^{(2)} \quad 5 \\
G_\kappa^{(3)}(ff) &= \left(f^\dagger \times \tilde{f} \right)_\kappa^{(3)} \quad 7, \\
G_\kappa^{(4)}(ff) &= \left(f^\dagger \times \tilde{f} \right)_\kappa^{(4)} \quad 9 \\
G_\kappa^{(5)}(ff) &= \left(f^\dagger \times \tilde{f} \right)_\kappa^{(5)} \quad 11 \\
G_\kappa^{(6)}(ff) &= \left(f^\dagger \times \tilde{f} \right)_\kappa^{(6)} \quad 13
\end{aligned}
\tag{9.171}
$$

for a total of 49 elements.

Subalgebra I: $u(7) \supset so(7) \supset g_2 \supset so(3) \supset so(2)$
The elements are:

$$so(7):$$
$$\left(f^\dagger \times \tilde{f}\right)^{(k)}_\kappa \quad k = 1,3,5 \quad 21$$

$$g_2;$$
$$\left(f^\dagger \times \tilde{f}\right)^{(k)}_\kappa \quad k = 1,5 \quad 14$$

$$so(3):$$
$$\left(f^\dagger \times \tilde{f}\right)^{(1)}_\kappa \quad\quad 3$$

$$so(2):$$
$$\left(f^\dagger \times \tilde{f}\right)^{(1)}_0 \quad\quad 1$$

$$(9.172)$$

Branching

The branching of representations $[N]$ of $u(7)$ is

$u(7)$ $[N] \equiv [N,0,0,0,0,0,0]$

$so(7)$ $\omega \equiv (\omega,0,0)$ $\omega = N, N-2, \ldots, 1$ or 0 (N = odd or even)

g_2 $\gamma \equiv (\gamma,0)$ Racah algorithm

$so(3)$ L Racah algorithm

$so(2)$ M_L $-L \le M_L \le +L$

$$(9.173)$$

The steps $so(7) \supset g_2$ and $g_2 \supset so(3)$ are non-canonical and involve the exceptional group g_2. The Racah algorithm gives Table 9.4. The algorithm can be found in Racah (1949). Note that for bosonic representations the algebra g_2 does not provide any new label and the multiplicities must be resolved by additional missing labels. (The algebra g_2 will be also discussed in Chap. 10 for fermionic systems for which it provides new labels.)

Table 9.4 Decomposition of totally symmetric representations of $u(7) \supset so(7)$ into representations of $so(3)$

$u(7) \quad \supset$	$so(7) \quad \supset$	$g_2 \quad \supset$	$so(3)$
N	ω	γ	L
0	0	0	0
1	1	1	3
2	2	2	$6, 4, 2$
	0	0	0
3	3	3	$9, 7, 6, 5, 4, 3, 1$
	1	1	3
4	4	4	$12, 10, 9, 8^2, 7, 6^2, 5^2, 4, 3^2, 2, 1$
	2	2	$6, 4, 2$
	0	0	0

9.8.2 The Triplet Chains

These chains can be constructed with two triplet vector bosons, $b^\dagger_{\rho,m}$ $(m = 0, \pm 1)$, $b^\dagger_{\lambda,m}$ $(m = 0, \pm 1)$ and a scalar boson $s^\dagger$, generically denoted $c^\dagger_\alpha$ $(\alpha = 1, \ldots, 7)$

$$l = 1: \quad b^\dagger_{\rho,m} \ (m = 0, \pm 1)$$

$$l = 1: \quad b^\dagger_{\lambda,m} \ (m = 0, \pm 1) \tag{9.174}$$

$$l = 0: \quad s^\dagger$$

together with the corresponding annihilation operators c_α $(\alpha = 1, \ldots, 7)$ (Bijker et al. 1994a). These chains are important in the study of the three-body problem in quantum mechanics. The subscript ρ, λ that distinguishes the two vector bosons is used to indicate that these vector bosons represent the second quantized form of the Jacobi variables ρ, λ that characterize the geometric configuration of a three-body system. The Jacobi variables are defined in terms of the coordinates of the three particles $\mathbf{r}_1, \mathbf{r}_2, \mathbf{r}_3$ as $\rho = \frac{1}{\sqrt{2}} (\mathbf{r}_1 - \mathbf{r}_2)$ and $\lambda = \frac{1}{\sqrt{6}} (\mathbf{r}_1 + \mathbf{r}_2 - 2\mathbf{r}_3)$. The bilinear products

$$G_{\alpha\alpha'} \doteq c^\dagger_\alpha c_{\alpha'} \quad (\alpha, \alpha' = 1, \ldots, 7) \tag{9.175}$$

generate $u(7)$. The basis states are written as

$$\mathcal{B}: \quad \frac{1}{\mathcal{N}} \left(b^\dagger_\rho\right)^{n_\rho} \left(b^\dagger_\lambda\right)^{n_\lambda} \left(s^\dagger\right)^{N - n_\rho - n_\lambda} |0\rangle, \tag{9.176}$$

where $\mathcal{N}$ is a normalization constant. The Racah form of $u(7)$ is obtained by introducing the operators

$$\tilde{b}_{\rho,m} = (-1)^{1-m} b_{\rho,-m}$$
$$\tilde{b}_{\lambda,m} = (-1)^{1-m} b_{\lambda,-m}$$
$$\tilde{s} = s \ . \tag{9.177}$$

It can be written as

$$\hat{n}_s = \left(s^\dagger \times \tilde{s}\right) \qquad\qquad 1$$

$$\hat{D}_{\rho,\mu} = \left(b_\rho^\dagger \times \tilde{s} - s^\dagger \times \tilde{b}_\rho\right)_\mu^{(1)} \qquad 3$$

$$\hat{D}_{\lambda,\mu} = \left(b_\lambda^\dagger \times \tilde{s} - s^\dagger \times \tilde{b}_\lambda\right)_\mu^{(1)} \qquad 3$$

$$\hat{A}_{\rho,\mu} = i\left(b_\rho^\dagger \times \tilde{s} + s^\dagger \times \tilde{b}_\rho\right)_\mu^{(1)} \qquad 3$$

$$\hat{A}_{\lambda,\mu} = i\left(b_\lambda^\dagger \times \tilde{s} + s^\dagger \times \tilde{b}_\lambda\right)_\mu^{(1)} \qquad 3 \tag{9.178}$$

$$\hat{G}_{S,\mu}^{(\ell)} = \left(b_\rho^\dagger \times \tilde{b}_\rho + b_\lambda^\dagger \times \tilde{b}_\lambda\right)_\mu^{(\ell)} \qquad 9$$

$$\hat{G}_{A,\mu}^{(\ell)} = i\left(b_\rho^\dagger \times \tilde{b}_\lambda - b_\lambda^\dagger \times \tilde{b}_\rho\right)_\mu^{(\ell)} \qquad 9$$

$$\hat{G}_{M\rho,\mu}^{(\ell)} = \left(b_\rho^\dagger \times \tilde{b}_\lambda + b_\lambda^\dagger \times \tilde{b}_\rho\right)_\mu^{(\ell)} \qquad 9$$

$$\hat{G}_{M\lambda,\mu}^{(\ell)} = \left(b_\rho^\dagger \times \tilde{b}_\rho - b_\lambda^\dagger \times \tilde{b}_\lambda\right)_\mu^{(\ell)} \qquad 9$$

with $\ell = 0, 1, 2$. This form is not the usual Racah form of the algebra in which the bilinear products are only angular momentum coupled. In applications to the many-body problem, especially in quantum chemistry, it is useful to assign to the creation and annihilation operators properties under transformations of a discrete group. The most important property is parity, where the transformation is inversion of the coordinates $\mathbf{r}_i \to -\mathbf{r}_i$. The two boson operators $b_\rho^\dagger$ and $b_\lambda^\dagger$ are assumed to be odd under parity, while the operator $s^\dagger$ is assumed to be even. (In general, in this chapter, the parity P of the boson operators $b_{l,m}^\dagger$ is assumed to be $(-1)^l$.) For applications to the three-body problem, it is convenient to assign to the elements of the algebra definite transformation properties under permutation of the three-bodies. The boson operators $s^\dagger, b_{\rho,\mu}^\dagger, b_{\lambda,m}^\dagger$ are assumed to transform under the transposition $P(12)$ as

$$P(12)\begin{pmatrix} s^\dagger \\ b_{\rho,m}^\dagger \\ b_{\lambda,m}^\dagger \end{pmatrix} = \begin{pmatrix} 1 & 0 & 0 \\ 0 & -1 & 0 \\ 0 & 0 & 1 \end{pmatrix}\begin{pmatrix} s^\dagger \\ b_{\rho,m}^\dagger \\ b_{\lambda,m}^\dagger \end{pmatrix}, \tag{9.179}$$

and under the cyclic permutation $P(123)$ as

$$P(123) \begin{pmatrix} s^\dagger \\ b^\dagger_{\rho,m} \\ b^\dagger_{\lambda,m} \end{pmatrix} = \begin{pmatrix} 1 & 0 & 0 \\ 0 & \cos(2\pi/3) & \sin(2\pi/3) \\ 0 & -\sin(2\pi/3) & \cos(2\pi/3) \end{pmatrix} \begin{pmatrix} s^\dagger \\ b^\dagger_{\rho,m} \\ b^\dagger_{\lambda,m} \end{pmatrix}. \tag{9.180}$$

The permutation group S_3 is a six element discrete group which is isomorphic to the dihedral group D_3. In order to characterize the transformation properties of the elements of the algebra (and of the states) one can use the label of either group, given by

$$S_3 \qquad D_3$$

$$S \equiv \square\,\square\,\square \quad A_1$$

$$M \equiv \begin{matrix} \square\,\square \\ \square \end{matrix} \quad E \,. \tag{9.181}$$

$$A \equiv \begin{matrix} \square \\ \square \\ \square \end{matrix} \quad A_2$$

This notation is used in (9.178). The construction of bosonic algebras with definite transformation properties under discrete groups is discussed in Bijker and Leviatan (1994b).

Subalgebras II

Since the number of elements of $u(7)$ is relatively large, there are several possible subalgebra chains, four of which are included here.

Subalgebra II.1

This is the chain

$$u(7) \supset u(6) \supset so(6) \supset so_\rho(3) \oplus so_\lambda(3) \supset so(3) \supset so(2). \tag{9.182}$$

States in this chain are characterized by the quantum numbers

$$\begin{vmatrix} u(7) \supset u(6) \supset so(6) \supset so_\rho(3) \oplus so_\lambda(3) \supset so(3) \supset so(2) \\ \downarrow \quad \downarrow \quad \downarrow \quad \downarrow \quad \downarrow \quad \downarrow \quad \downarrow \\ N \quad n \quad \gamma \quad L_\rho \quad L_\lambda \quad L \quad M_L \end{vmatrix} \tag{9.183}$$

with branching

$$
\begin{array}{llll}
u(7) & N \equiv [N,0,0,0,0,0,0] & \\
u(6) & n \equiv [n,0,0,0,0,0] & n = N, N-1, \ldots, 0 \\
so(6) & \gamma \equiv (\gamma,0,0) & \gamma = n, n-2, \ldots, 1 \text{ or } 0 \ (n\text{=odd or even}) \\
so_\rho(3) \oplus so_\lambda(3) & L_\rho, L_\lambda & \text{Algorithm 5} \\
so(3) & L & |L_\rho - L_\lambda| \le L \le |L_\rho + L_\lambda| \\
so(2) & M_L & -L \le M_L \le +L
\end{array}
$$

(9.184)

Algorithm 5. The values of L_ρ and L_λ are obtained by partitioning γ as

$$ \gamma = 2\nu + L_\rho + L_\lambda \quad \text{with } \nu = 0, 1, \ldots \tag{9.185} $$

Subalgebra II.2
   This is the chain

$$ u(7) \supset so(7) \supset so(6) \supset so_\rho(3) \oplus so_\lambda(3) \supset so(3) \supset so(2). \tag{9.186} $$

States in this chain are characterized by

$$
\left|
\begin{array}{ccccccc}
u(7) \supset & so(7) \supset & so(6) \supset & so_\rho(3) \oplus so_\lambda(3) \supset & so(3) \supset & so(2) \\
\downarrow & \downarrow & \downarrow & \downarrow \quad\quad\quad \downarrow & \downarrow & \downarrow \\
N & \omega & \gamma & L_\rho \quad\quad\quad L_\lambda & L & M_L
\end{array}
\right\rangle,
\tag{9.187}
$$

with branching

$$
\begin{array}{llll}
u(7) & N \equiv [N,0,0,0,0,0,0] & \\
so(7) & \omega \equiv (\omega,0,0) & \omega = N, N-2, \ldots, 1 \text{ or } 0 \\
so(6) & \gamma \equiv (\gamma,0,0) & \gamma = \omega, \omega-1, \ldots, 0 \\
so_\rho(3) \oplus so_\lambda(3) & L_\rho, L_\lambda & \text{Algorithm 5} \\
so(3) & L & |L_\rho - L_\lambda| \le L \le |L_\rho + L_\lambda| \\
so(2) & M_L & -L \le M_L \le +L
\end{array}
$$

(9.188)

Subalgebras III

Two other subalgebra chains have been used extensively in hadronic physics (three-quark system) (Bijker and Leviatan 1994b).
   Subalgebra III.1
   This is the chain

$$u(7) \supset u_\rho(3) \oplus u_\lambda(4) \supset u_\rho(3) \oplus u_\lambda(3) \supset so_\rho(3) \oplus so_\lambda(3) \supset so(3) \supset so(2).$$
(9.189)

States in this chain are characterized by the quantum numbers

$$\left| \begin{array}{l} u(7) \supset u_\rho(3) \oplus u_\lambda(4) \supset u_\rho(3) \oplus u_\lambda(3) \supset so_\rho(3) \oplus so_\lambda(3) \\ \quad\downarrow \qquad\qquad\qquad\qquad\quad \downarrow \quad\ \downarrow \qquad\qquad \downarrow \quad\ \downarrow \\ \quad N \qquad\qquad\qquad\qquad\quad n_\rho \quad n_\lambda \qquad\qquad L_\rho \quad L_\lambda \\ \\ \qquad\qquad\quad \supset so(3) \supset so(2) \\ \qquad\qquad\qquad \downarrow \qquad \downarrow \\ \qquad\qquad\qquad L \qquad M_L \end{array} \right\rangle,$$
(9.190)

with branching

$$u(7) \qquad N \equiv [N,0,0,0,0,0,0]$$

$$u_\rho(3) \oplus u_\lambda(3) \qquad \begin{array}{l} n_\rho \equiv (n_\rho,0,0) \qquad n_\rho = 0,1,\ldots,N \\ n_\lambda \equiv (n_\lambda,0,0) \qquad n_\lambda = 0,1,\ldots,N-n_\rho \end{array}$$

$$so_\rho(3) \oplus so_\lambda(3) \qquad \begin{array}{l} L_\rho \qquad L_\rho = n_\rho, n_\rho - 2, \ldots, 1 \text{ or } 0 \ . \\ L_\lambda \qquad L_\lambda = n_\lambda, n_\lambda - 2, \ldots, 1 \text{ or } 0 \end{array}$$

$$so(3) \qquad L \qquad \left| L_\rho - L_\lambda \right| \le L \le \left| L_\rho + L_\lambda \right|$$

$$so(2) \qquad M_L \qquad -L \le M_L \le +L$$
(9.191)

Subalgebra III.2
This is the chain

$$u(7) \supset u_\rho(3) \oplus u_\lambda(4) \supset u_\rho(3) \oplus so_\lambda(4) \supset so_\rho(3) \oplus so_\lambda(3) \supset so(3) \supset so(2).$$
(9.192)

States in this chain are characterized by the quantum numbers

$$
\left| \begin{array}{l}
u(7) \supset u_\rho(3) \oplus u_\lambda(4) \supset u_\rho(3) \oplus so_\lambda(4) \supset so_\rho(3) \oplus so_\lambda(3) \\
\quad\downarrow \qquad\qquad\qquad\qquad \downarrow \qquad \downarrow \qquad\qquad\quad \downarrow \qquad \downarrow \\
\quad N \qquad\qquad\qquad\qquad\; n_\rho \qquad \omega \qquad\qquad L_\rho \qquad L_\lambda \\
\\
\qquad\qquad\qquad\qquad \supset so(3) \supset so(2) \\
\qquad\qquad\qquad\qquad\quad \downarrow \qquad \downarrow \\
\qquad\qquad\qquad\qquad\quad L \qquad M_L
\end{array} \right\rangle ,
\tag{9.193}
$$

with branching

$$u(7) \qquad N \equiv [N, 0, 0, 0, 0, 0, 0]$$

$$
\begin{array}{lll}
u_\rho(3) \oplus so_\lambda(4) & n_\rho \equiv (n_\rho, 0, 0) & n_\rho = 0, 1, \ldots, N \\
& \omega \equiv (\omega, 0) & \omega = N - n_\rho, N - n_\rho - 2, \ldots, 1 \text{ or } 0 \\
\\
so_\rho(3) \oplus so_\lambda(3) & L_\rho & L_\rho = n_\rho, n_\rho - 2, \ldots, 1 \text{ or } 0 \\
& L_\lambda & L_\lambda = 0, 1, \ldots, \omega \\
\\
so(3) & L & \left| L_\rho - L_\lambda \right| \le L \le \left| L_\rho + L_\lambda \right| \\
\\
so(2) & M_L & -L \le M_L \le +L
\end{array}
$$

$$\tag{9.194}$$

Lattice of Algebras

The lattice of algebras for the chains discussed above is

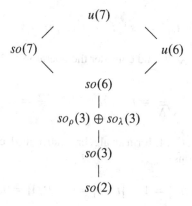

and

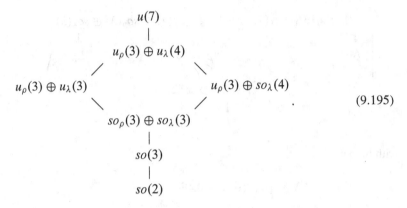

$$(9.195)$$

Note that the two lattices merge at the level of $so_\rho(3) \oplus so_\lambda(3)$ (Iachello 1995).

9.9   Contractions of Bosonic Algebras

The bosonic realizations of the Lie algebras $u(n)$ of (9.2) with commutation relations (9.3) are well suited for constructing contracted Lie algebras. These contracted algebras are particularly useful in quantum mechanics.

9.9.1   The Heisenberg Algebra $h(2)$

We begin by considering the algebra $u(2)$ constructed with two boson operators $(s,t) \equiv b_\alpha(\alpha = 1,2)$ satisfying the commutation relations (9.1). The elements of $u(2)$ can be written as

$$\hat{F}_- = s^\dagger t, \quad \hat{F}_+ = t^\dagger s, \quad \hat{n}_t = t^\dagger t, \quad \hat{n}_s = s^\dagger s. \qquad (9.196)$$

Now replace s and $s^\dagger$ by $\sqrt{N}$, and consider the operators

$$\frac{\hat{F}_-}{\sqrt{N}} = t, \quad \frac{\hat{F}_+}{\sqrt{N}} = t^\dagger, \quad \hat{n}_t = t^\dagger t, \quad \frac{\hat{n}_s}{N} = 1. \qquad (9.197)$$

The set of operators $t, t^\dagger, t^\dagger t, 1$, form an algebra, called the Heisenberg algebra $h(2)$, with commutation relations

$$[t, t^\dagger] = 1, \quad [t^\dagger, 1] = 0, \quad [t, 1] = 0,$$
$$[t, t^\dagger t] = t, \quad [t^\dagger, t^\dagger t] = -t^\dagger. \qquad (9.198)$$

The algebra $h(2)$ has a subalgebra $u(1)$ composed of the single element $t^\dagger t$,

$$h(2) \supset u(1), \tag{9.199}$$

and it is the contracted algebra of $u(2)$

$$u(2) \to_c h(2). \tag{9.200}$$

The three elements $t, t^\dagger$ and 1 also form an algebra, called the quantum mechanical algebra in one dimension, $q(1)$. The origin of this name can be seen by introducing a coordinate, x, and momentum, $p_x = \frac{1}{i}\frac{d}{dx}$, realization of the boson operators

$$t = \frac{1}{\sqrt{2}}(x + ip_x), \quad t^\dagger = \frac{1}{\sqrt{2}}(x - ip_x), \tag{9.201}$$

or, conversely,

$$x = \frac{1}{\sqrt{2}}(t + t^\dagger), \quad p_x = \frac{1}{\sqrt{2}}i(t^\dagger - t). \tag{9.202}$$

The elements of the Heisenberg algebra $h(2)$ in terms of coordinate and momentum are

$$x, \quad ip_x, \quad 1, \quad x^2 + p_x^2. \tag{9.203}$$

Similarly, the elements of the quantum mechanical algebra $q(1)$ are

$$x, ip_x, 1 \quad (\text{or} \quad x, p_x, i) \tag{9.204}$$

whose commutation relations are those of the basic commutation relations of quantum mechanics, (1.3),

$$[x, ip_x] = -1 \quad (\text{or} \quad [x, p_x] = i). \tag{9.205}$$

9.9.2 The Heisenberg Algebra $h(n)$, $n = $ Even

These algebras can be easily constructed by introducing boson operators in the Racah form and splitting them into a scalar boson, s, and a tensor boson operator, b_l ($l = 0, 1, 2, \ldots$). The case $l = 0$ is the case discussed in the previous subsection. We construct here explicitly the case of $u(4)$, which leads to the contracted algebra $h(4)$ and to the quantum mechanical algebra $q(3)$. We begin by introducing the boson operators (9.112) with $l = 0$ (σ) and $l = 1$ ($\pi_\mu, \mu = 0, \pm1, \pm2$). Here Greek letters are used not to confuse boson operators with the momentum p. The elements of $u(4)$ in Racah form are given in (9.113). We now replace $\sigma, \sigma^\dagger$ by $\sqrt{N}$ obtaining the operators

$$G_0^{(0)} (\sigma\sigma) / N = 1$$
$$G_0^{(0)} (\pi\pi) = \left(\pi^\dagger \times \tilde{\pi}\right)_0^{(0)}$$
$$G_\kappa^{(1)} (\pi\pi) = \left(\pi^\dagger \times \tilde{\pi}\right)_\kappa^{(1)}$$
$$G_\kappa^{(2)} (\pi\pi) = \left(\pi^\dagger \times \tilde{\pi}\right)_\kappa^{(2)}$$
$$G_\kappa^{(1)} (\pi\sigma) / \sqrt{N} = \left(\pi^\dagger\right)_\kappa^{(1)}$$
$$G_\kappa^{(1)} (\sigma\pi) / \sqrt{N} = (\tilde{\pi})_\kappa^{(1)} . \tag{9.206}$$

The 16 operators

$$1, \quad \pi^\dagger, \quad \tilde{\pi}, \quad \left(\pi^\dagger \times \tilde{\pi}\right)^{(k)} (k = 0, 1, 2) \tag{9.207}$$

form a contracted Lie algebra, called the Heisenberg algebra $h(4)$. This algebra has the subalgebra chain

$$h(4) \supset u(3) \supset so(3) \supset so(2). \tag{9.208}$$

Instead of the operators $1, \pi^\dagger, \tilde{\pi}, \left(\pi^\dagger \times \tilde{\pi}\right)^{(k)}$, one can also consider the set of operators $1, \left(\pi^\dagger + \tilde{\pi}\right), i\left(\pi^\dagger - \tilde{\pi}\right), \left(\pi^\dagger \times \tilde{\pi}\right)^{(k)}$, that is, one could do the contraction on the operators $\hat{D}$ and $\hat{D}'$ of (9.124).

Introducing coordinates, $\mathbf{r}$, and momenta, $\mathbf{p}$, in three-dimensions,

$$\pi = \frac{1}{\sqrt{2}} (\mathbf{r} + i\mathbf{p}), \quad \pi^\dagger = \frac{1}{\sqrt{2}} (\mathbf{r} - i\mathbf{p}), \tag{9.209}$$

the algebra $h(4)$ can be rewritten as composed of

$$\mathbf{r}, \mathbf{p}, 1, (\mathbf{r}^2 + \mathbf{p}^2), (\mathbf{r} \times \mathbf{p})^{(1)} = \hat{L}, (\mathbf{r} \times \mathbf{r} + \mathbf{p} \times \mathbf{p})^{(2)} = \hat{Q}. \tag{9.210}$$

From this form, one can see that among the elements of the Heisenberg algebra $h(4)$ there are the coordinates, the momenta, the angular momentum and the quadrupole tensor. The quantum mechanical algebra $q(3)$ is simply composed of $\mathbf{r}, \mathbf{p}, 1$.

A similar procedure can be used to construct the Heisenberg algebra $h(6)$ and the associated quantum mechanical algebra in five-dimensions, $q(5)$. One starts from the boson operators $l = 0 (s)$ and $l = 2 (d_\mu, \mu = 0, \pm 1, \pm 2)$ of (9.135). The Heisenberg algebra $h(6)$ is composed of the 36 operators

$$1, \quad d^\dagger, \quad \tilde{d}, \quad \left(d^\dagger \times \tilde{d}\right)^{(k)} \quad (k = 0, 1, 2, 3, 4). \tag{9.211}$$

Introducing coordinates and momenta (q_μ, p_μ) $(\mu = 0, \pm 1, \pm 2)$, one obtains the quantum mechanical algebra composed of $q_\mu, p_\mu, 1$.

Chapter 10
Fermion Realizations

10.1 Fermion Operators

In Chap. 9 a realization of Lie algebras in terms of boson operators has been given. These operators have been used either in connection with coordinates and momenta, as in harmonic oscillator problems, see Chap. 9, Examples 1, 2 and 3, or as operators which create or annihilate particles with integer values of the angular momentum $l = 0, 1, \ldots$. In the 1920s, it became evident that particles exist with half-integer values of the angular momentum, $j = \frac{1}{2}, \frac{3}{2}, \ldots$. These particles are called fermions. For applications to physics, it is of interest to consider realizations of Lie algebras in terms of fermion operators. In view of the pervasive presence in physics of particles with half-integer spin, most notably electrons and nucleons, fermion realizations have become an important part of Lie algebraic methods. These will be discussed in this chapter.

We begin by introducing an operation called anticommutator, denoted by a curly bracket, $\{,\}$. The anticommutator of two quantities X, Y is

$$\{X, Y\} = XY + YX. \tag{10.1}$$

The anticommutator is sometimes denoted by $[,]_+$. In this book, the curly bracket notation will be used. We then introduce fermion creation, $a_i^\dagger$, and annihilation, a_i, operators, satisfying anticommutation relations,

$$\left\{a_i, a_{i'}^\dagger\right\} = \delta_{ii'} \quad ; \quad \{a_i, a_{i'}\} = \left\{a_i^\dagger, a_{i'}^\dagger\right\} = 0. \tag{10.2}$$

These operators are the key ingredient in the construction.

© Springer-Verlag Berlin Heidelberg 2015
F. Iachello, *Lie Algebras and Applications*, Lecture Notes in Physics 891,
DOI 10.1007/978-3-662-44494-8_10

10.2   Lie Algebras Constructed with Fermion Operators

Lie algebras can be constructed from bilinear products of fermion creation and annihilation operators

$$g \doteq \quad A_{ik} = a_i^\dagger a_k \quad (i, k = 1, \dots, n), \tag{10.3}$$

as in the previous case of boson operators. The elements A_{ik} satisfy the commutation relations of the unitary algebra $u(n)$ [and $gl(n)$]

$$[A_{ik}, A_{st}] = A_{it} \delta_{ks} - A_{sk} \delta_{it}. \tag{10.4}$$

A basis can be constructed by acting with the creation operators on a vacuum state $|0\rangle$. This basis will be denoted by $\mathcal{F}$.

$$\mathcal{F}: \quad \frac{1}{\mathcal{N}} \, a_i^\dagger a_{i'}^\dagger \dots |0\rangle \,. \tag{10.5}$$

In contrast with the case of boson operators which generate totally symmetric representations, the irreducible representations of $u(n)$ generated by acting with fermion operators on $|0\rangle$ are the totally antisymmetric representations, with Young tableau

$$\left. \begin{array}{c} \square \\ \square \\ \vdots \\ \square \end{array} \right\} N_F. \tag{10.6}$$

For $u(n)$ the Young tableau has n entries. The totally antisymmetric representations have N_F entries 1 and $n - N_F$ entries 0

$$\{N_F\} \equiv \overbrace{[\underbrace{1, 1, \dots, 1}_{N_F}, 0, \dots, 0]}^{n}. \tag{10.7}$$

The short-hand notation $\{N_F\}$ is often used to denote these representations. The zeros are usually not written, except for the identity representation $[0, 0, \dots, 0] \equiv [0]$. The basis $\mathcal{F}$ is often called a Fermi–Dirac basis. Since any Lie algebra is a subalgebra of $gl(n)$ it can be constructed with bilinear products of fermion operators.

10.3 Racah Form

While in the case of bosons both the uncoupled and coupled (Racah) form of the
Lie algebra have been extensively used, in the case of fermions only the Racah form
has been to a large extent used, and will be considered in this book. The Racah form
of the Lie algebra can be obtained by introducing fermion operators that transform
as representations $|j, m\rangle$ of $spin(3) \supset spin(2)$, with j =half-integer. We use here
$spin(3)$ and $spin(2)$ instead of $so(3)$ and $so(2)$, since we need to consider explicitly
spinor representations. The corresponding creation and annihilation operators will
be denoted by

$$
\begin{aligned}
a^\dagger_{j,m} \quad & m = \pm\tfrac{1}{2}, \pm\tfrac{3}{2}, \ldots, \pm j \\
a_{j,m} \quad & m = \pm\tfrac{1}{2}, \pm\tfrac{3}{2}, \ldots, \pm j
\end{aligned}
\tag{10.8}
$$

These operators satisfy anticommutation relations

$$
\begin{aligned}
\{a_{j,m}, a^\dagger_{j',m'}\} &= \delta_{jj'} \delta_{mm'} \\
\{a_{j,m}, a_{j',m'}\} &= \{a^\dagger_{j,m}, a^\dagger_{j',m'}\} = 0
\end{aligned}
\tag{10.9}
$$

In constructing the Lie algebra it is convenient to introduce the operators

$$
\tilde{a}_{j,m} = (-1)^{j-m} a_{j,-m}
\tag{10.10}
$$

that transform as tensors under $spin(3) \supset spin(2)$. (An alternative definition is
with a phase $(-1)^{j+m}$. This alternative definition introduces minus signs in some
formulas, but it does not alter the algebraic structure. Operators with physical
meaning constructed with creation and annihilation operators must be defined
accordingly. This problem does not arise for Racah boson realizations, since l and
m are integers.)

The Racah form of $u(n)$ is

$$
\begin{aligned}
g \doteq \quad A^{(\lambda)}_\mu(j, j') &= [a^\dagger_j \times \tilde{a}_{j'}]^{(\lambda)}_\mu \\
&= \sum_{m,m'} \langle jmj'm'|\lambda\mu\rangle a^\dagger_{j,m}\tilde{a}_{j',m'} ,
\end{aligned}
\tag{10.11}
$$

with $|j + j'| \geq \lambda \geq |j - j'|$ and $n = \sum_i (2j_i + 1)$. The commutation relations
are

$$
\left[A^{(\lambda)}_\mu(j, j'), A^{(\lambda')}_{\mu'}(j'', j''')\right] = -\sum_{\lambda''\mu''} \left[(2\lambda + 1)(2\lambda' + 1)\right]^{1/2} \langle \lambda\mu\lambda'\mu' \mid \lambda''\mu''\rangle
$$

$$\times \left[(-)^{\lambda''+j+j'''} \begin{Bmatrix} \lambda & \lambda' & \lambda'' \\ j''' & j & j' \end{Bmatrix} \times \delta_{j'j''} A_{\mu''}^{(\lambda'')}(j, j''') \right.$$

$$\left. - (-)^{\lambda+\lambda'+j'+j''} \begin{Bmatrix} \lambda & \lambda' & \lambda'' \\ j'' & j' & j \end{Bmatrix} \delta_{jj'''} A_{\mu''}^{(\lambda'')}(j'', j') \right]. \quad (10.12)$$

Note that, since j is half-integer, one can construct with fermion operators that transform as representations of $spin(3) \supset spin(2)$ only unitary algebras in an even number of dimensions, a peculiarity of these realizations.

10.4   The Algebras $u(2j + 1)$

Consider a single value of j. There are $(2j + 1)$ values of m. Thus $n = 2j+1$. With this single value of j, we can construct the algebra $u(2j + 1)$. There are $(2j + 1)^2$ elements in the algebra

$$A_{\mu}^{(\lambda)}(j, j) = [a_j^\dagger \times \tilde{a}_j]_\mu^{(\lambda)}. \quad (10.13)$$

To the algebra $u(2j + 1)$ we can now apply the general procedure of constructing subalgebras.

10.4.1   Subalgebra Chain Containing Spin(3)

A generic subalgebra chain (called a classification scheme) for $u(2j+1)$ constructed with fermion operators can be obtained as follows:

1. Exclude the element with $\lambda = 0$, $\mu = 0$; this gives $su(2j + 1)$.
2. Retain only terms with $\lambda =$odd; this gives the Lie algebra $sp(2j + 1, C) \equiv sp(2j + 1)$. Note that this situation is different from that of Lie algebras constructed in terms of boson operators, since in that case, retaining terms with $\lambda =$odd generates the orthogonal Lie algebras, $so(2l + 1)$.
3. Retain the term with $\lambda = 1$; this gives the algebra $spin(3)$.
4. Retain the term with $\lambda = 1, \mu = 0$; this gives the algebra $spin(2)$. A generic subalgebra chain for single j is thus

$$u(2j + 1) \supset su(2j + 1) \supset sp(2j + 1) \supset \ldots \supset spin(3) \supset spin(2). \quad (10.14)$$

Dots have been inserted between $sp(2j + 1)$ and $spin(3)$, since, for large j, there may be intermediate steps (Flowers 1952).

10.4.2 The Algebras $u(2)$ and $su(2)$: Spinors

Because of the pervasive presence in physics of particles with angular momentum $\frac{1}{2}$, it is of great interest to construct explicitly the Lie algebra $u(2)$ in terms of fermion operators with $j = \frac{1}{2}$, called spinors. The fermion creation operators are written as

$$a^\dagger_{\frac{1}{2},+\frac{1}{2}} \quad , \quad a^\dagger_{\frac{1}{2},-\frac{1}{2}}. \tag{10.15}$$

An alternative notation, often found in books in condensed matter physics, is $a^\dagger_\uparrow$, $a^\dagger_\downarrow$, called spin up, and spin down notation.

The 4 elements of $u(2)$ are

$$A^{(1)}_\mu(\tfrac{1}{2}, \tfrac{1}{2}) = \left[a^\dagger_{\frac{1}{2}} \times \tilde{a}_{\frac{1}{2}}\right]^{(1)}_\mu \quad 3$$

$$A^{(0)}_0(\tfrac{1}{2}, \tfrac{1}{2}) = \left[a^\dagger_{\frac{1}{2}} \times \tilde{a}_{\frac{1}{2}}\right]^{(0)}_0 \quad 1 \tag{10.16}$$

By deleting the element $A^{(0)}_0$ one obtains the subalgebra $su(2) \sim sp(2) \sim spin(3)$. By considering only the element $A^{(1)}_0$, one obtains the subalgebra $spin(2)$. The basis is

$$\left| \begin{array}{ccc} u(2) & \supset su(2) & \supset spin(2) \\ \downarrow & & \downarrow \\ N_F & & M_J \end{array} \right\rangle. \tag{10.17}$$

Note that, for totally antisymmetric representations, no new quantum number is introduced when going from $u(2)$ to $su(2) \sim sp(2)$. The elements of the Lie algebra $u(2)$ constructed with spin $\frac{1}{2}$ fermion operators have a straightforward physical meaning. They are the total spin operator, $\hat{S}$, and the number operator for fermions, $\hat{N}_F$,

$$\hat{S}_\mu = -\sqrt{\tfrac{1}{2}}\left[a^\dagger_{1/2} \times \tilde{a}_{1/2}\right]^{(1)}_\mu$$

$$\hat{N}_F = -\sqrt{2}\left[a^\dagger_{1/2} \times \tilde{a}_{1/2}\right]^{(0)}_0 \tag{10.18}$$

(The minus sign arises from the choice of phases in (10.10).) The only states of $u(2)$ that can be constructed with fermion operators are

$$\square \equiv [1] \quad N_F = 1$$

$$\begin{array}{c}\square\\\square\end{array} \equiv [1, 1] \quad N_F = 2 \tag{10.19}$$

Table 10.1 Classification of antisymmetric states of $u(2)$

$j = 1/2$	$u(2)$	$su(2)$
	$[\lambda_1, \lambda_2]$	J
	$[0]$	0
	$[1]$	$1/2$
	$[1, 1]$	0

in addition to the vacuum $|0\rangle$ that transforms as the identity representation $[0]$. This property is known to physicists as Pauli principle. The classification of antisymmetric states is rewritten, for comparison with the subsequent sections, as in Table 10.1.

The algebra of $u(2)$ constructed with fermion operators has had many applications in physics, and hence many different notations have been used to label the states. Antisymmetric states of $u(2)$ have been labeled either by $\{N_F\}$ or by $[\lambda_1, \lambda_2]$. The number of fermions, N_F, is related to the labels $[\lambda_1, \lambda_2]$ by

$$N_F = \lambda_1 + \lambda_2. \tag{10.20}$$

Also, in this case, in the general classification scheme of (10.14), all steps, except the last one, coincide, since $su(2) \sim sp(2) \sim spin(3)$. Finally, either $u(2)$ or $su(2)$ can be used to classify the states, and the use of both the labels $[\lambda_1, \lambda_2]$ of $u(2)$ and the label J of $su(2)$ is redundant.

10.4.3   The Algebra $u(4)$

Another case of considerable interest is the case of $j = \frac{3}{2}, n = 4$. The 16 elements of the algebra are

$$A_\mu^{(3)}(\tfrac{3}{2}, \tfrac{3}{2}) = \left[a_{3/2}^\dagger \times \tilde{a}_{3/2} \right]_\mu^{(3)} \quad 7$$

$$A_\mu^{(2)}(\tfrac{3}{2}, \tfrac{3}{2}) = \left[a_{3/2}^\dagger \times \tilde{a}_{3/2} \right]_\mu^{(2)} \quad 5$$

$$A_\mu^{(1)}(\tfrac{3}{2}, \tfrac{3}{2}) = \left[a_{3/2}^\dagger \times \tilde{a}_{3/2} \right]_\mu^{(1)} \quad 3 \tag{10.21}$$

$$A_0^{(0)}(\tfrac{3}{2}, \tfrac{3}{2}) = \left[a_{3/2}^\dagger \times \tilde{a}_{3/2} \right]_0^{(0)} \quad 1$$

By deleting the element $A_0^{(0)}(\tfrac{3}{2}, \tfrac{3}{2})$ one obtains the subalgebra $su(4)$. By retaining the 10 elements with λ =odd

$$A_\mu^{(3)}\left(\tfrac{3}{2}, \tfrac{3}{2}\right) = \left[a_{3/2}^\dagger \times \tilde{a}_{3/2}\right]_\mu^{(3)} \quad 7$$

$$A_\mu^{(1)}\left(\tfrac{3}{2}, \tfrac{3}{2}\right) = \left[a_{3/2}^\dagger \times \tilde{a}_{3/2}\right]_\mu^{(1)} \quad 3$$
(10.22)

one obtains the algebra of $sp(4)$. By retaining only the elements with $\lambda = 1$

$$A_\mu^{(1)}\left(\tfrac{3}{2}, \tfrac{3}{2}\right) = \left[a_{3/2}^\dagger \times \tilde{a}_{3/2}\right]_\mu^{(1)} \quad 3$$
(10.23)

one obtains the algebra of $spin(3) \sim su(2)$. Finally by retaining only

$$A_0^{(1)}\left(\tfrac{3}{2}, \tfrac{3}{2}\right) = \left[a_{3/2}^\dagger \times \tilde{a}_{3/2}\right]_0^{(1)} \quad 1$$
(10.24)

one obtains $spin(2)$. A basis for fermions with spin $\tfrac{3}{2}$ is then

$$\left| \begin{array}{cccc} u(4) \supset & sp(4) & \supset su(2) \supset & spin(2) \\ \downarrow & \downarrow & \downarrow & \downarrow \\ N_F & (n_1, n_2) & J & M_J \end{array} \right\rangle.$$
(10.25)

The values of n_1, n_2 are restricted by the branching rules and only one quantum number is actually needed. The allowed states are

$$\square \equiv [1] \qquad N_F = 1$$

$$\begin{array}{c}\square\\\square\end{array} \equiv [1, 1] \qquad N_F = 2$$

$$\begin{array}{c}\square\\\square\\\square\end{array} \equiv [1, 1, 1] \quad N_F = 3.$$
(10.26)

$$\begin{array}{c}\square\\\square\\\square\\\square\end{array} \equiv [1, 1, 1, 1] \quad N_F = 4$$

The classification of antisymmetric states is given in Table 10.2.
The number N_F is here

Table 10.2 Classification of antisymmetric states of $u(4)$

	$u(4)$	$sp(4)$	$su(2)$
$j = 3/2$	$[\lambda_1, \lambda_2, \lambda_3, \lambda_4]$	(n_1, n_2)	J
	$[0]$	$(0,0)$	0
	$[1]$	$(1,0)$	$3/2$
	$[1,1]$	$(0,0)$	0
		$(1,1)$	2
	$[1,1,1]$	$(1,0)$	$3/2$
	$[1,1,1,1]$	$(0,0)$	0

$$N_F = \sum_{i=1}^{4} \lambda_i. \tag{10.27}$$

10.4.4   The Algebra $u(6)$

This algebra can be constructed with $j = \frac{5}{2}$. The 36 elements have the form

$$A_\mu^{(5)}\left(\tfrac{5}{2}, \tfrac{5}{2}\right) = \left[a_{5/2}^\dagger \times \tilde{a}_{5/2}\right]_\mu^{(5)} \quad 11$$

$$A_\mu^{(4)}\left(\tfrac{5}{2}, \tfrac{5}{2}\right) = \left[a_{5/2}^\dagger \times \tilde{a}_{5/2}\right]_\mu^{(4)} \quad 9$$

$$A_\mu^{(3)}\left(\tfrac{5}{2}, \tfrac{5}{2}\right) = \left[a_{5/2}^\dagger \times \tilde{a}_{5/2}\right]_\mu^{(3)} \quad 7$$

$$A_\mu^{(2)}\left(\tfrac{5}{2}, \tfrac{5}{2}\right) = \left[a_{5/2}^\dagger \times \tilde{a}_{5/2}\right]_\mu^{(2)} \quad 5 \tag{10.28}$$

$$A_\mu^{(1)}\left(\tfrac{5}{2}, \tfrac{5}{2}\right) = \left[a_{5/2}^\dagger \times \tilde{a}_{5/2}\right]_\mu^{(1)} \quad 3$$

$$A_0^{(0)}\left(\tfrac{5}{2}, \tfrac{5}{2}\right) = \left[a_{5/2}^\dagger \times \tilde{a}_{5/2}\right]_0^{(0)} \quad 1$$

Subalgebras of $u(6)$ can be constructed as in (10.14).

The basis $\mathcal{F}$ can be labeled by

$$\left| \begin{matrix} u(6) \supset & sp(6) & \supset su(2) \supset & spin(2) \\ \downarrow & \downarrow & \downarrow & \downarrow \\ N_F & (n_1, n_2, n_3) & J & M_J \end{matrix} \right). \tag{10.29}$$

Table 10.3 Classification of antisymmetric states of $u(6)$

$j = 5/2$	$u(6)$ $[\lambda_1, \lambda_2, \lambda_3, \lambda_4, \lambda_5, \lambda_6]$	$sp(6)$ (n_1, n_2, n_3)	$su(2)$ J
	$[0]$	$(0, 0, 0)$	0
	$[1]$	$(1, 0, 0)$	$5/2$
	$[1, 1]$	$(0, 0, 0)$	0
		$(1, 1, 0)$	$2, 4$
	$[1, 1, 1]$	$(1, 0, 0)$	$5/2$
		$(1, 1, 1)$	$3/2, 9/2$
	$[1, 1, 1, 1]$	$(0, 0, 0)$	0
		$(1, 1, 0)$	$2, 4$
	$[1, 1, 1, 1, 1]$	$(1, 0, 0)$	$5/2$
	$[1, 1, 1, 1, 1, 1]$	$(0, 0, 0)$	0

The representations (n_1, n_2, n_3) of $sp(6)$ are restricted and only one quantum number is needed to classify uniquely the states. The classification scheme is given in Table 10.3. The representations of $u(6)$ are labelled here by their Young tableau $[\lambda_1, \lambda_2, \lambda_3, \lambda_4, \lambda_5, \lambda_6]$. The number of fermions N_F is

$$N_F = \sum_{i=1}^{6} \lambda_i. \tag{10.30}$$

Algebras with $j > \frac{5}{2}$ and their classification scheme can be constructed in a similar way.

10.4.5   Branchings of $u(2j + 1)$

For all representations of the previous subsections, one needs the branchings of $u(2j + 1)$ into $sp(2j + 1)$ and of $sp(2j + 1)$ into $su(2)$. These branchings are discussed in detail and tabulated for $j \leq 7/2$, in Hamermesh (1962).

10.5   The Algebra $u\left(\sum_i (2j_i + 1)\right)$

A generalization of the construction given in the previous section, is the case when there are several values of j. These situations occur in atomic and nuclear physics where they are called mixed configurations. The construction is straightforward and it produces the Lie algebra $u(n)$, with $n = \sum_i (2j_i + 1)$. Note, however, once more, that with fermion operators that transform as representations of $spin(3) \supset spin(2)$ it is possible to construct only unitary algebras in an even number of dimensions.

Example 1. The algebras $u(6)$ and $u(12)$

In the study of the spectroscopy of nuclei, one encounters situations in which the values of j are $j = \frac{1}{2}, \frac{3}{2}$ or $j = \frac{1}{2}, \frac{3}{2}, \frac{5}{2}$. The algebras constructed with these values are $u(6)$ and $u(12)$ (Iachello and van Isacker 1991).

10.6   Internal Degrees of Freedom (Different Spaces)

10.6.1   The Algebras $u(4)$ and $su(4)$

One often encounters in physics particles with internal degrees of freedom. These objects transform as representation $|j, m\rangle$ under $spin(3) \supset spin(2)$, and as representations of some internal symmetry group, G. A particularly interesting case is that of protons and neutrons. These particles transform as $j = \frac{1}{2}$ under $spin(3)$ and as $t = \frac{1}{2}$ under another group called $isospin(3)$. The fermion creation operators are, in a double index notation:

$$
\begin{array}{ll}
a^\dagger_{\frac{1}{2},+\frac{1}{2},\frac{1}{2},+\frac{1}{2}} & p\uparrow \\[4pt]
a^\dagger_{\frac{1}{2},-\frac{1}{2},\frac{1}{2},+\frac{1}{2}} & p\downarrow \\[4pt]
a^\dagger_{\frac{1}{2},+\frac{1}{2},\frac{1}{2},-\frac{1}{2}} & n\uparrow \\[4pt]
a^\dagger_{\frac{1}{2},-\frac{1}{2},\frac{1}{2},-\frac{1}{2}} & n\downarrow
\end{array}
\tag{10.31}
$$

generically denoted $a^\dagger_{s,m_s,t,m_t}$. The annihilation operators are a_{s,m_s,t,m_t}. The 16 bilinear products of creation and annihilation operators form an algebra, called Wigner $u(4)$. It is convenient to introduce the adjoint operators

$$
\tilde{a}_{s,m_s,t,m_t} = (-)^{s-m_s+t-m_t}\, a_{s,-m_s,t,-m_t},
\tag{10.32}
$$

that transform as tensors under spin and isospin rotations. The Racah form of Wigner $u(4)$ is obtained in terms of double tensors

$$
\left[a^\dagger_{\frac{1}{2},\frac{1}{2}} \times \tilde{a}_{\frac{1}{2},\frac{1}{2}} \right]^{(S,T)}_{(M_S,M_T)} = \sum_{\substack{m,m' \\ m_t,m_{t'}}} \langle \frac{1}{2},m_s,\frac{1}{2},m'_s \mid S,M_S \rangle
$$

$$
\times \langle \frac{1}{2},m_t,\frac{1}{2},m_{t'} \mid T,M_T \rangle a^\dagger_{\frac{1}{2},m_s,\frac{1}{2},m_t} \tilde{a}_{\frac{1}{2},m'_s,\frac{1}{2},m'_t}
\tag{10.33}
$$

with $S = 0, 1$ and $T = 0, 1$.

Subalgebra Chain

Although a given algebra can be decomposed into several subalgebra chains, as shown in the previous chapter for bosonic realizations, in practice physical considerations dictate what subalgebra chains are of interest. For algebras with internal degrees of freedom, the construction of subalgebra chains splits into two categories: different spaces and same spaces. For different spaces, the internal degrees of freedom are separated from the outset and not combined at any stage, while for same spaces, they are combined at some stage into a single algebra, whose elements are the sum of the elements of the two algebras. This procedure is elucidated in the paragraphs below.

In the case of Wigner $u(4)$, spin and isospin live on different spaces, the physical space and a fictitious isotopic spin space. It is assumed, on physical grounds, that they cannot be combined, and the appropriate chain is $u(4) \supset su(4) \supset su_S(2) \oplus su_T(2) \supset spin_S(2) \oplus spin_T(2)$. Here $su_S(2) \sim spin(3)$, $su_T(2) \sim isospin(3)$ and the subscript S, T has been added to distinguish spin from isospin. The basis $\mathcal{F}$ for Wigner $u(4)$ can then be written as

$$\left| \begin{array}{cccc} u(4) & \supset su_S(2) \oplus su_T(2) & \supset spin_S(2) \oplus spin_T(2) \\ \downarrow & \downarrow & \downarrow \\ [\lambda_1, \lambda_2, \lambda_3, \lambda_4] & S, T & M_S, M_T \end{array} \right). \quad (10.34)$$

The full notation for the representations of $u(4)$ has been restored and the intermediate step $su(4)$ has been omitted, since no new quantum number appears. For a given representation of $u(4)$, the branching problem can be solved using the techniques of Chap. 6. The chain is non-canonical and requires a building-up process. It has become customary to introduce another notation, called Wigner notation. In this notation, one first goes from $u(4)$ to $su(4)$, using the rules of Chap. 6. The representations of $su(4)$ are labeled by $\left[\lambda_1', \lambda_2', \lambda_3'\right]$ with

$$\lambda_1' = \lambda_1 - \lambda_4, \lambda_2' = \lambda_2 - \lambda_4, \lambda_3' = \lambda_3 - \lambda_4. \quad (10.35)$$

The Wigner quantum numbers (P, P', P'') are defined as

$$(P, P', P'') = \left(\frac{\lambda_1' + \lambda_2' - \lambda_3'}{2}, \frac{\lambda_1' - \lambda_2' + \lambda_3'}{2}, \frac{\lambda_1' - \lambda_2' - \lambda_3'}{2} \right). \quad (10.36)$$

The branching of representations of $su(4)$ into representations of $su_S(2) \oplus su_T(2)$ can then be constructed (Hamermesh 1962). A portion of this branching is given in Table 10.4. In the last column the dimension of the representations of $su(4)$ is shown.

10.6.2 The Algebras $u(6)$ and $su(6)$

In this case, particles still transform as $j = \frac{1}{2}$ under $spin(3)$, but they now transform as the fundamental representation [1] of an internal group $su_F(3)$, called the flavor

Table 10.4 Branching $su(4) \supset su_S(2) \oplus su_T(2)$

$su(4)$ (P, P', P'')	$\supset$	$su_S(2) \oplus su_T(2)$ (S, T)	$\dim(P, P', P'')$
$(0, 0, 0)$		$(0, 0)$	1
$(\frac{1}{2}, \frac{1}{2}, \frac{1}{2})$		$(\frac{1}{2}, \frac{1}{2})$	4
$(1, 1, 1)$		$(0, 0), (1, 1)$	10
$(\frac{3}{2}, \frac{3}{2}, \frac{3}{2})$		$(\frac{1}{2}, \frac{1}{2}), (\frac{3}{2}, \frac{3}{2})$	20

group. The creation operators are denoted by $a^\dagger_{u\uparrow}, a^\dagger_{d\uparrow}, a^\dagger_{s\uparrow}, a^\dagger_{u\downarrow}, a^\dagger_{d\downarrow}, a^\dagger_{s\downarrow}$. The index u, d, s stands for u, d, s quarks. The bilinear product of creation and annihilation operators form a $u(6)$ algebra, called the Gürsey–Radicati algebra.

Subalgebra Chain

The breaking of $u(6)$ first proceeds to $su_F(3) \oplus su_S(2)$. The algebra $su(3)$ can be broken in various ways. The breaking appropriate to the physical situation described by the Gürsey–Radicati algebra is $su_F(3) \supset su_T(2) \oplus u_Y(1)$, where T denotes the isospin as before, and $u_Y(1)$ is an Abelian algebra called hypercharge (Y) algebra. The subalgebra chain is thus $u(6) \supset su(6) \supset su_F(3) \oplus su_S(2) \supset su_T(2) \oplus u_Y(1) \oplus su_S(2) \supset spin_T(2) \oplus u_Y(1) \oplus spin_S(2)$. The basis is labelled by

$$
\left| \begin{array}{ccccccc}
u(6) \supset & su_F(3) & \oplus su_S(2) \supset & su_T(2) & \oplus u_Y(1) & \oplus su_S(2) \\
\downarrow & \downarrow & \downarrow & \downarrow & \downarrow \\
[\lambda] & [\mu_1, \mu_2] & S & T & Y \\
\\
& \supset spin_T(2) & \oplus u_Y(1) & \oplus spin_S(2) \\
& \downarrow & & \downarrow \\
& T_3 & & S_3
\end{array} \right\rangle . \qquad (10.37)
$$

Here $[\lambda] = [\lambda_1, \lambda_2, \lambda_3, \lambda_4, \lambda_5, \lambda_6]$. It has become customary to label the representations of $su_F(3)$ not by their Young labels $[\mu_1, \mu_2]$ but by the dimension of the representations, $\dim[0, 0] = 1$, $\dim[2, 1] = 8$, etc. and the representations of $su_S(2)$ by their dimensions $\dim S = 2S + 1$, put as a superscript, that is $^2 8$ denotes the representation $[2, 1]$, $S = \frac{1}{2}$ of $su_F(3) \oplus su_s(2)$. The branching of representations of $u(6)$ into representations of $su_F(3) \oplus su_S(2)$ can be constructed (Gürsey and Radicati 1964). A portion of this branching is given in Table 10.5. The conversion between this notation and the standard notation for $su(3)$ representations used in the preceding chapters is $1 \equiv [0, 0]$, $3 \equiv [1, 0]$, $\bar{3} \equiv [1, 1]$, $6 \equiv [2, 0]$, $8 \equiv [2, 1]$, $10 \equiv [3, 0]$. The general formula for the dimension of the representations of $u(n)$ is given in Chap. 6, Sect. 6.12. It is worth noting that with fermions with internal degrees of freedom, it is possible to construct algebras in both even and odd number of dimensions, in the present case $su_F(3)$.

Table 10.5 Branching $su(6) \supset su_F(3) \oplus su_T(2)$

$su(6)$ $[\lambda_1, \lambda_2, \lambda_3, \lambda_4, \lambda_5]$	$\supset$	$su_F(3) \oplus su_S(2)$ $(\dim [\mu_1, \mu_2], \dim S)$	$\dim [\lambda]$
$[0, 0, 0, 0, 0]$		$(1, 1)$	1
$[1, 0, 0, 0, 0]$		$(3, 2)$	6
$[1, 1, 1, 1, 1]$		$(\bar{3}, 2)$	6
$[1, 1, 0, 0, 0]$		$(\bar{3}, 3), (6, 1)$	15
$[1, 1, 1, 0, 0]$		$(8, 2), (1, 4)$	20
$[2, 0, 0, 0, 0]$		$(6, 3), (\bar{3}, 1)$	21
$[2, 1, 1, 1, 1]$		$(8, 3), (8, 1), (1, 3)$	35
$[3, 0, 0, 0, 0]$		$(10, 4), (8, 2)$	56
$[2, 1, 0, 0, 0]$		$(10, 2), (8, 4), (8, 2), (1, 2)$	70

10.7 Internal Degrees of Freedom (Same Space)

10.7.1 The Algebra $u((2l + 1)(2s + 1))$: L-S Coupling

Another case of interest is that of particles with both orbital, l =integer, and spin, s =half-integer, angular momentum. (The spin angular momentum is the internal degree of freedom.) The corresponding creation and annihilation operators are

$$a^\dagger_{l,m_l,s,m_s}$$
$$\tilde{a}_{l,m_l,s,m_s} = (-1)^{l-m_l+s-m_s} \, a_{l,-m_l,s,-m_s} . \tag{10.38}$$

The bilinear products

$$a^\dagger_{l,m_l,s,m_s} \tilde{a}_{l,m'_l,s,m'_s} \tag{10.39}$$

generate the Lie algebra $u((2l + 1)(2s + 1))$.

Racah Form

The Racah form of the orbital-spin algebra is constructed with double tensors as in the preceding section,

$$\left[a^\dagger_{l,s} \times \tilde{a}_{l,s} \right]^{(L,S)}_{M_L,M_S} = \sum_{\substack{m_s,m'_s \\ m_l,m'_l}} \langle l, m_l, l, m'_l \mid L, M_L \rangle \langle s, m_s, s, m'_s \mid S, M_S \rangle$$

$$\times a^\dagger_{l,m_l,s,m_s} \tilde{a}_{l,m'_l,s,m'_s} . \tag{10.40}$$

The classification scheme for this algebra is

$$u((2l+1)(2s+1)) \supset u_L(2l+1) \oplus su_S(2s+1) \supset \cdots$$
$$\supset so_L(3) \oplus su_S(2) \supset spin_J(3) \supset spin_J(2), \qquad (10.41)$$

called L-S coupling. Since orbital and spin angular momentum are assumed to act on the same space, they can be combined. Here $spin_J(3)$ denotes the algebra whose elements are the sum of the elements of $so_L(3)$ and $su_S(2)$, sometimes called the diagonal algebra. For large l, some intermediate steps are needed between $u_L(2l+1)$ and $so_L(3)$ and the problem of missing labels becomes particularly acute since the representations of $u_L(2l+1)$ that appear in (10.41) are not just one-row representations as in the case of bosons. For this reason, dots have been inserted in (10.41) between $u_L(2l+1)$ and $so_L(3)$.

The representations of $u((2l+1)(2s+1))$ that describe fermions are the totally antisymmetric representations $\{N_F\}$ of (10.7) (one-column representations). Racah showed that, in order to obtain totally antisymmetric representations of $u((2l+1)(2s+1))$, the representations of $u_L(2l+1)$ and $su_S(2s+1)$ must be *dual* (sometimes called *conjugate*), that is obtained from each other by interchanging columns with rows. Particularly interesting is the case of $s = \frac{1}{2}$ that describes electrons in atoms and nucleons in nuclei. In this case, the algebra is $u(2(2l+1)) \supset u_L(2l+1) \oplus su_S(2)$. The representations of $su_S(2)$ are characterized by the quantum numbers S, M_S, and thus described by a Young tableau with two rows of length $(N/2) + S$ and $(N/2) - S$. The representations of $u_L(2l+1)$ which arise in the reduction of representations $\{N_F\}$ of $u(2(2l+1))$ must then be described by a Young tableau with two columns of these lengths.

Example 2. p electrons in atoms

In this case, $l = 1$, leading to the Lie algebra $u(6)$. The classification scheme is

$$u(6) \supset u_L(3) \oplus su_S(2) \supset so_L(3) \oplus su_S(2) \supset spin_J(3) \supset spin_J(2). \qquad (10.42)$$

The branching of the representations of $u_L(3)$ into those of $so_L(3)$ has been discussed in Chap. 6, Sect. 6.14. No additional subalgebra is needed.

Example 3. f electrons in atoms

In this case, $l = 3$, and the appropriate algebra is $u(14)$. This algebra has 196 elements. The classification scheme of (10.41) is

$$u(14) \supset u_L(7) \oplus su_S(2) \supset \ldots \supset so_L(3) \oplus su_S(2) \supset spin_J(3) \supset spin_J(2). \qquad (10.43)$$

The classification scheme for the step $u_L(7) \supset so_L(3)$ is incomplete. A complete classification scheme was found by Racah and it is

$$u_L(7) \supset so_L(7) \supset g_2 \supset so_L(3) \supset so_L(2). \tag{10.44}$$

This classification scheme occupies a special role in Lie algebraic methods in physics since it was the first case in which an exceptional algebra, g_2, appeared. The elements of the algebras in (10.44) can be written easily in the double tensor notation of (10.40). They are

$u(7)$:
$$\left[a^\dagger \times \tilde{a}\right]^{(L,0)}_{M_L,0} \quad L = 0,1,2,3,4,5,6 \quad 49$$

$su(7)$:
$$\left[a^\dagger \times \tilde{a}\right]^{(L,0)}_{M_L,0} \quad L = 1,2,3,4,5,6 \quad 48$$

$so(7)$:
$$\left[a^\dagger \times \tilde{a}\right]^{(L,0)}_{M_L,0} \qquad L = 1,3,5 \qquad 21$$

g_2 :
$$\left[a^\dagger \times \tilde{a}\right]^{(L,0)}_{M_L,0} \qquad L = 1,5 \qquad 14 \tag{10.45}$$

$so(3)$:
$$\left[a^\dagger \times \tilde{a}\right]^{(L,0)}_{M_L,0} \qquad L = 1 \qquad 3$$

$so(2)$:
$$\left[a^\dagger \times \tilde{a}\right]^{(1,0)}_{M_L,0} \qquad\qquad\qquad 1$$

where the indices $l = 3, s = \frac{1}{2}$ have been omitted. States are characterized by

$$\left| \begin{array}{ccccc} u(7) & \supset & so(7) & \supset & g_2 & \supset so(3) \supset so(2) \\ \downarrow & & \downarrow & & \downarrow & \downarrow \quad\quad \downarrow \\ [\lambda_1,\ldots,\lambda_7] & & [\mu_1,\mu_2,\mu_3] & & [\gamma_1,\gamma_2] & L \quad\quad M_L \end{array} \right\rangle. \tag{10.46}$$

For the application described in this example, the representations of $u(14)$ are characterized by the number of electrons N_F. The representations of $su_S(2)$ are characterized by the total spin S. For each N_F and S the representations of $u(7)$ are two-column representations with length $(N_F/2) + S$ and $(N_F/2) - S$. Tables of reduction for $u(7) \supset so(7)$, $so(7) \supset g_2$, $g_2 \supset so(3)$ can be found in Racah. Note that the reduction $u(7) \supset so(7)$ for fermionic representations is different from that of bosonic representations given in Chap. 9. The same remark applies for the reduction $so(7) \supset g_2$ and $g_2 \supset so(3)$. The reduction $so(7) \supset g_2$ is given in Table 10.6. The reduction $g_2 \supset so(3)$ is given in Table 10.7. To complete the study of this chain, one needs also to construct the Casimir invariants and their eigenvalues. The construction is given in Racah. The general expression for the eigenvalues in the representation $[\gamma_1, \gamma_2]$ are

$$\langle C_2(g_2)\rangle = \left(\gamma_1^2 + \gamma_1\gamma_2 + \gamma_2^2 + 5\gamma_1 + 4\gamma_2\right)/12. \tag{10.47}$$

(The factor of 12 comes from the definition of $C_2(g_2)$.) A detailed account is given in Racah (1949). Also, note again that with fermions with internal degrees of freedom

Table 10.6 Reduction $so(7) \supset g_2$

$so(7)$	$\supset$	g_2
$[\mu_1, \mu_2, \mu_3]$		$[\gamma_1, \gamma_2]$
$[0, 0, 0]$		$[0, 0]$
$[1, 0, 0]$		$[1, 0]$
$[1, 1, 0]$		$[1, 0], [1, 1]$
$[2, 0, 0]$		$[2, 0]$
$[1, 1, 1]$		$[0, 0], [1, 0], [2, 0]$
$[2, 1, 0]$		$[1, 1], [2, 0], [2, 1]$
$[2, 1, 1]$		$[1, 0], [1, 1], [2, 0], [2, 1], [3, 0]$
$[2, 2, 0]$		$[2, 0], [2, 1], [2, 2]$
$[2, 2, 1]$		$[1, 0], [1, 1], [2, 0], [2, 1], [3, 0], [3, 1]$
$[2, 2, 2]$		$[0, 0], [1, 0], [2, 0], [3, 0], [4, 0]$

Table 10.7 Reduction $g_2 \supset so(3)$

g_2	$\supset$	$so(3)$
$[\gamma_1, \gamma_2]$		L
$[0, 0]$		0
$[1, 0]$		3
$[1, 1]$		$1, 5$
$[2, 0]$		$2, 4, 6$
$[2, 1]$		$2, 3, 4, 5, 7, 8$
$[3, 0]$		$1, 3, 4, 5, 6, 7, 9$
$[2, 2]$		$0, 2, 4, 5, 6, 8, 10$
$[3, 1]$		$1, 2, 3^2, 4, 5^2, 6^2, 7^2, 8, 9, 10, 11$
$[4, 0]$		$0, 2, 3, 4^2, 5, 6^2, 7, 8^2, 9, 10, 12$

it is possible to construct unitary algebras in odd number of dimensions, here $u(7)$. Finally, $L - S$ coupling plays a major role in the spectroscopy of atoms.

10.7.2 The Algebra $u\left(\sum_j (2j + 1)\right)$: j-j Coupling

The orbital and spin angular momenta can be coupled from the outset. The corresponding creation and annihilation fermion operators are

$$a^\dagger_{l,s,j,m_j}; \quad \tilde{a}_{l,s,j,m_j} = (-1)^{j-m_j} a_{l,s,j,-m_j}, \tag{10.48}$$

with $|l + s| \geq j \geq |l - s|$. The bilinear products

$$a^\dagger_{l,s,j,m_j} \tilde{a}_{l,s,j',m'_j}$$

form the Lie algebra $u(n)$ with $n = \sum_j (2j + 1)$, as in Sect. 10.5. The Racah form of the algebra and its branchings can be constructed (Iachello and van Isacker 1991). Again here the case of interest is that in which $s = \frac{1}{2}$.

Example 4. p nucleons in nuclei

In this case, $l = 1$, and $j = \frac{1}{2}, \frac{3}{2}$, leading to the Lie algebra $u(6)$. The classification scheme is

$$u(6) \supset sp(6) \supset spin_J(3) \supset spin_J(2). \tag{10.49}$$

j-j coupling plays a major role in the spectroscopy of nuclei.

10.7.3   The Algebra $u\left(\left(\sum_l (2l + 1)\right)(2s + 1)\right)$: Mixed L-S Configurations

In this case the orbital angular momentum l takes several values $l_1, l_2, \ldots$. The creation and annihilation operators are still given by (10.48), but the elements of the Lie algebra are the bilinear products

$$a^\dagger_{l,m_l,s,m_s} \tilde{a}_{l',m'_l,s,m'_s} \tag{10.50}$$

where l, l' take the values $l_1, l_2, \ldots$. A particularly important case described by this algebra is that of fermions with spin $s = \frac{1}{2}$ moving in a harmonic oscillator potential in three dimensions. As discussed in Chap. 9, Sect. 9.5.1, for $N \geq 2$ states of the three dimensional harmonic oscillator are degenerate with respect to the angular momentum. For $N = 2$ the values of the orbital angular momentum are $l = 0, 2$, for $N = 3$ they are $l = 1, 3$, etc. The algebra that describes fermions in a given degenerate shell (a given value of N), called Elliott algebra, is $u\left(\left(\sum_l (2l + 1)\right)(2s + 1)\right)$ (Elliott 1958).

Example 5. s-d nucleons in nuclei

In this case $l = 0, 2$ leading to the Lie algebra $u(12)$. The classification scheme is

$$u(12) \supset u_L(6) \oplus su_S(2) \supset su_L(3) \oplus su_S(2) \supset so_L(3) \oplus su_S(2)$$

$$\supset spin_J(3) \supset spin_J(2). \tag{10.51}$$

Mixed L-S configurations play a major role in the spectroscopy of light nuclei.

Chapter 11
Differential Realizations

11.1 Differential Operators

Lie algebras can also be constructed with differential operators acting on a space of
derivable functions $f(x)$ of the coordinate x. The basic commutation relations are

$$\left[x, \frac{d}{dx}\right] = x\frac{d}{dx} - \frac{d}{dx}x = -1. \tag{11.1}$$

11.2 Unitary Algebras $u(n)$

The bilinear products of coordinates, x_i $(i = 1, \ldots, n)$, and their derivatives,
$\frac{\partial}{\partial x_j}$ $(j = 1, \ldots, n)$ generate $u(n)$

$$u(n) \doteq x_i \frac{\partial}{\partial x_j}. \tag{11.2}$$

Introducing the double index notation of the previous chapters

$$X_{ik} = x_i \frac{\partial}{\partial x_k} \quad (i, k = 1, \ldots, n), \tag{11.3}$$

one obtains the commutation relations

$$[X_{ik}, X_{mn}] = \delta_{km} X_{in} - \delta_{in} X_{mk}. \tag{11.4}$$

For $su(n)$, one subtracts $\sum_j X_{jj}$ from the diagonal elements

© Springer-Verlag Berlin Heidelberg 2015
F. Iachello, *Lie Algebras and Applications*, Lecture Notes in Physics 891,
DOI 10.1007/978-3-662-44494-8_11

$$X'_{ii} = X_{ii} - \frac{1}{n} \sum_j X_{jj}. \tag{11.5}$$

and deletes X'_{nn}. (The n elements X'_{ii} are not linear independent since $\sum_i X'_{ii} = 0$.)
The $n^2 - 1$ elements $X_{ik}(i \neq k = 1, \ldots, n)$ and $X'_{ii}(i = 1, \ldots, n - 1)$ generate
$su(n)$.

Example 1. Differential realization of $u(2)$

Introduce two coordinates x, y. The differential realization is

$$X_{11} = x\frac{\partial}{\partial x}, X_{12} = x\frac{\partial}{\partial y}, X_{21} = y\frac{\partial}{\partial x}, X_{22} = y\frac{\partial}{\partial y}. \tag{11.6}$$

Example 2. Differential realization of $su(2)$

From the preceding example, one obtains

$$X'_{11} = \frac{1}{2}\left(x\frac{\partial}{\partial x} - y\frac{\partial}{\partial y}\right), X_{12} = x\frac{\partial}{\partial y}, X_{21} = y\frac{\partial}{\partial x}. \tag{11.7}$$

11.3   Orthogonal Algebras $so(n)$

Differential realization are often used to construct orthogonal Lie algebras. Introduce n real coordinates, $x_1, x_2, .., x_n$. A construction of the Lie algebra $so(n)$ is

$$so(n) \doteq x_i\frac{\partial}{\partial x_j} - x_j\frac{\partial}{\partial x_i}. \tag{11.8}$$

Introducing the notation

$$L_{ij} = x_i\frac{\partial}{\partial x_j} - x_j\frac{\partial}{\partial x_i}, \quad i < j = 1, \ldots, n. \tag{11.9}$$

one obtains the commutation relations

$$\left[L_{ij}, L_{kl}\right] = \delta_{jk}L_{il} + \delta_{il}L_{jk} - \delta_{jl}L_{ik} - \delta_{ik}L_{jl}. \tag{11.10}$$

The $\frac{n(n-1)}{2}$ elements L_{ij} are called angular momentum operators.

Example 3. The Lie algebra so(2)

We introduce here two coordinates, x, y, as in part (a) of Fig. 11.1. The single
element of $so(2)$ is

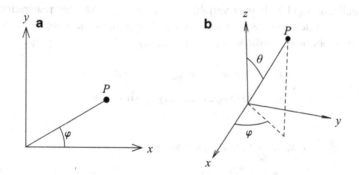

Fig. 11.1   Coordinates for differential realizations of (**a**) $so(2)$ and (**b**) $so(3)$

$$X_1 = y\frac{\partial}{\partial x} - x\frac{\partial}{\partial y}. \tag{11.11}$$

This element acts on functions $f(x, y)$.

Example 4.   The Lie algebra so(3)

We introduce here three coordinates, x, y, z, as in part (b) of Fig. 11.1. The three elements of $so(3)$ are, in a single index notation,

$$X_1 = z\frac{\partial}{\partial y} - y\frac{\partial}{\partial z}$$

$$X_2 = x\frac{\partial}{\partial z} - z\frac{\partial}{\partial x}$$

$$X_3 = y\frac{\partial}{\partial x} - x\frac{\partial}{\partial y} \tag{11.12}$$

acting of $f(x, y, z)$. The elements satisfy the commutation relations

$$[X_1, X_2] = X_3 \; ; \; [X_2, X_3] = X_1 \; ; \; [X_3, X_1] = X_2. \tag{11.13}$$

Example 5.   The Lie algebra so(4)

We introduce here four coordinates, x, y, z, t. The six elements of $so(4)$ are

$$A_1 = z\frac{\partial}{\partial y} - y\frac{\partial}{\partial z}, A_2 = x\frac{\partial}{\partial z} - z\frac{\partial}{\partial x}, A_3 = y\frac{\partial}{\partial x} - x\frac{\partial}{\partial y},$$

$$B_1 = x\frac{\partial}{\partial t} - t\frac{\partial}{\partial z}, B_2 = y\frac{\partial}{\partial t} - t\frac{\partial}{\partial y}, B_3 = z\frac{\partial}{\partial t} - t\frac{\partial}{\partial z}. \tag{11.14}$$

acting on $f(x, y, z, t)$.

The realization (11.8) is in n variables $x_1, x_2, \ldots, x_n$. Another realization is in terms of $n-1$ angle variables $\varphi_1, \varphi_2, \ldots, \varphi_{n-1}$. This realization is obtained by introducing hyperspherical coordinates, $(x_1, x_2, \ldots, x_n) \rightarrow (r, \varphi_{n-1}, \varphi_{n-2}, \varphi_{n-3}, \ldots, \varphi_1)$

$$x_1 = r \sin \varphi_{n-1} \sin \varphi_{n-2} \ldots \sin \varphi_2 \sin \varphi_1$$

$$x_2 = r \sin \varphi_{n-1} \sin \varphi_{n-2} \ldots \sin \varphi_2 \cos \varphi_1$$

$$\ldots$$

$$x_{n-1} = r \sin \varphi_{n-1} \cos \varphi_{n-2}$$

$$x_n = r \cos \varphi_{n-1} \tag{11.15}$$

and setting $r = 1$ (realization on the unit n dimensional sphere.)

Example 6. so(2) in polar coordinates

Introducing

$$\begin{aligned} x &= r \cos \varphi \\ y &= r \sin \varphi \end{aligned} \tag{11.16}$$

one has

$$X_1 = \frac{\partial}{\partial \varphi}. \tag{11.17}$$

Example 7. so(3) in spherical coordinates

Introducing

$$\begin{aligned} x &= r \sin \vartheta \cos \varphi \\ y &= r \sin \vartheta \sin \varphi \, , \\ z &= r \cos \vartheta \end{aligned} \tag{11.18}$$

one has

$$\begin{aligned} X_1 &= \cos \varphi \frac{\partial}{\partial \vartheta} + \sin \varphi \, ctg \, \vartheta \frac{\partial}{\partial \varphi} \\ X_2 &= -\sin \varphi \frac{\partial}{\partial \vartheta} - \cos \varphi \, ctg \, \vartheta \frac{\partial}{\partial \varphi} \, . \\ X_3 &= \frac{\partial}{\partial \varphi} \end{aligned} \tag{11.19}$$

Example 8. so(4) in hyperspherical coordinates

One needs here three angles, ϑ, φ, ψ. The corresponding form of the Lie algebra is called Pauli form.

11.3.1 Casimir Operators: Laplace-Beltrami Form

All operators are constructed in terms of differential operators. The differential form of the Casimir operators is often referred to as Laplace-Beltrami form.

Example 9. The Laplace-Beltrami form of so(2)

We have here trivially

$$C_1(so(2)) = \frac{\partial}{\partial\varphi}, \quad C_2(so(2)) = \frac{\partial^2}{\partial\varphi^2}. \tag{11.20}$$

Example 10. The Laplace-Beltrami form of so(3)

We have here

$$C_2(so(3)) = \frac{1}{\sin^2\vartheta}\frac{\partial^2}{\partial\varphi^2} + \frac{1}{\sin\vartheta}\frac{\partial}{\partial\vartheta}(\sin\vartheta\frac{\partial}{\partial\vartheta}). \tag{11.21}$$

11.3.2 Basis for the Representations

One also needs to construct the basis for the representations. In the realization in terms of angles, the basis is constructed using the canonical chain

$$so(n) \supset so(n-1) \supset \ldots \supset so(2). \tag{11.22}$$

Example 11. Basis for so(3)

In this case one seeks simultaneous eigenfunctions of $C_2(so(3))$ and $C_2(so(2))$

$$C_2(so(2))\,\Phi(\varphi) = -m^2\,\Phi(\varphi)$$
$$C_2(so(3))\,\Theta(\vartheta) = -l(l+1)\,\Theta(\vartheta). \tag{11.23}$$

The eigenfunctions are the spherical harmonics

$$Y_l^m(\vartheta,\varphi) = \sqrt{\frac{2l+1}{4\pi}\frac{(l-m)!}{(l+m)!}}(-1)^m\,e^{im\varphi}\,P_l^m(\cos\vartheta). \tag{11.24}$$

It has become customary in physical applications to introduce the elements L_x, L_y, L_z obtained from those previously given by multiplication with $(1/i)$

$$\begin{cases} L_x = \frac{1}{i}\left(\cos\varphi\frac{\partial}{\partial\vartheta} - \sin\varphi ctg\vartheta \frac{\partial}{\partial\varphi}\right) \\[2ex] L_y = \frac{1}{i}(-\sin\varphi\frac{\partial}{\partial\vartheta} - \cos\varphi\, ctg\,\vartheta\, \frac{\partial}{\partial\varphi})\ . \\[2ex] L_z = \frac{1}{i}\frac{\partial}{\partial\varphi} \end{cases} \qquad (11.25)$$

From these one can also construct the elements in Cartan-Weyl form, $L_z, L_\pm = L_x \pm iL_y$. The action of those elements on the basis is

$$\begin{cases} L_z\, Y_l^m(\vartheta,\varphi) = m\, Y_l^m(\vartheta,\varphi) \\[2ex] L_+\, Y_l^m(\vartheta,\varphi) = \sqrt{(l-m)(l+m+1)}\, Y_l^{m+1}(\vartheta,\varphi)\ . \\[2ex] L_-\, Y_l^m(\vartheta,\varphi) = \sqrt{(l+m)(l-m+1)}\, Y_l^{m-1}(\vartheta,\varphi) \end{cases} \qquad (11.26)$$

The abstract notation is

$$\left| \begin{matrix} so(3) \supset so(2) \\ \downarrow \qquad \downarrow \\ l \qquad\quad m \end{matrix} \right\rangle \equiv Y_l^m(\vartheta,\varphi). \qquad (11.27)$$

Note that with differential realizations it is possible to construct only tensor representations of the orthogonal algebras. The construction of spinor representations requires the introduction of another mathematical framework, either that of spinors (Cartan 1966) or that of Grassmann variables (Berezin 1987).

11.4   Orthogonal Algebras $so(n, m)$

Although the discussion in the preceding chapters has been devoted to compact Lie algebras, it is of interest to consider here non-compact Lie algebras, $so(n, m)$. A construction of the Lie algebra $so(n, m)$ is

$$so(n, m) \doteq \delta_i x_i \frac{\partial}{\partial x_j} - \delta_j x_j \frac{\partial}{\partial x_i}, \qquad (11.28)$$

where

$$\delta_i = \begin{cases} 1 & i = 1,\ldots,n \\ -1 & i = n+1,\ldots,n+m \end{cases} \qquad (11.29)$$

Example 12. The Lorentz algebra $so(3, 1)$

Introducing coordinates x, y, z, t, the elements are

$$A_1 = x\frac{\partial}{\partial y} - y\frac{\partial}{\partial x}, A_2 = y\frac{\partial}{\partial z} - z\frac{\partial}{\partial y}, A_3 = z\frac{\partial}{\partial x} - x\frac{\partial}{\partial z},$$

$$B_1 = x\frac{\partial}{\partial t} + t\frac{\partial}{\partial x}, B_2 = y\frac{\partial}{\partial t} + t\frac{\partial}{\partial y}, B_3 = z\frac{\partial}{\partial t} + t\frac{\partial}{\partial z}. \tag{11.30}$$

11.5 Symplectic Algebras $sp(2n)$

A differential realization of $sp(2n)$ is obtained by introducing $2n$ coordinates divided into $x_1, \ldots, x_n; x_1', \ldots, x_n'$. A construction of $sp(2n)$ is

$$
\begin{array}{lll}
x_i \frac{\partial}{\partial x_i'} & i = 1, \ldots, n & n \\
x_i' \frac{\partial}{\partial x_i} & i = 1, \ldots, n & n \\
x_i \frac{\partial}{\partial x_j} - x_j' \frac{\partial}{\partial x_i'} & i, j = 1, \ldots, n & n^2 \\
x_i \frac{\partial}{\partial x_j'} + x_j \frac{\partial}{\partial x_i'} & i < j = 1, \ldots, n & \frac{n(n-1)}{2} \\
x_i' \frac{\partial}{\partial x_j} + x_j' \frac{\partial}{\partial x_i} & i < j = 1, \ldots, n & \frac{n(n-1)}{2}
\end{array}
\tag{11.31}
$$

The number of elements is written to their right. There are in total $n(2n + 1)$ elements.

Example 13. The Lie algebra $sp(2)$

Introducing coordinates x_1, x_1' one has

$$X_1 = x_1\frac{\partial}{\partial x_1'}, X_2 = x_1'\frac{\partial}{\partial x_1}, X_3 = x_1\frac{\partial}{\partial x_1} - x_1'\frac{\partial}{\partial x_1'}. \tag{11.32}$$

All the differential realizations discussed above are real forms (Chen 1989).

Chapter 12
Matrix Realizations

12.1 Matrices

Lie algebras can be constructed with $n \times n$ square matrices

$$A = \begin{pmatrix} \cdots \\ \cdots \\ \cdots \end{pmatrix}$$

(12.1)

acting on the right on column vectors with n rows

$$\chi = \begin{pmatrix} \cdot \\ \cdot \\ \cdot \end{pmatrix} ,$$

(12.2)

and on the left on row vectors with n columns

$$\tilde{\chi} = \begin{pmatrix} \cdots \end{pmatrix} .$$

(12.3)

Properties of matrices have been given in Chap. 3.

© Springer-Verlag Berlin Heidelberg 2015
F. Iachello, *Lie Algebras and Applications*, Lecture Notes in Physics 891,
DOI 10.1007/978-3-662-44494-8_12

12.2   Unitary Algebras $u(n)$

The matrices

$$E_{km} = \begin{pmatrix} & & & 0 & & \\ & & & 0 & & \\ 0 & 0 & 0 & 1 & 0 & 0 \\ & & & 0 & & \\ & & & 0 & & \\ & & & 0 & & \end{pmatrix} \tag{12.4}$$

with 1 in the k-th row and m-th column and zero otherwise, satisfy the commutation relations of $u(n)$ [and $gl(n)$]

$$[E_{km}, E_{st}] = \delta_{sm} E_{kt} - \delta_{kt} E_{sm}. \tag{12.5}$$

For $su(n)$, one subtracts the unit matrix, I, divided by n,

$$I = \begin{pmatrix} 1 & & & & 0 \\ & 1 & & & \\ & & \cdots & & \\ & & & 1 & \\ 0 & & & & 1 \end{pmatrix} \tag{12.6}$$

from E_{kk}

$$E'_{kk} = E_{kk} - \frac{1}{n} I \tag{12.7}$$

and deletes E'_{nn}. The remaining $n^2 - 1$ matrices are the elements of $su(n)$. This procedure provides real forms of $u(n)$ and $su(n)$.

Example 1. Matrix realization of $u(2)$

A real matrix realization of $u(2)$ is

$$E_{11} = \begin{pmatrix} 1 & 0 \\ 0 & 0 \end{pmatrix}, \quad E_{12} = \begin{pmatrix} 0 & 1 \\ 0 & 0 \end{pmatrix},$$

$$E_{21} = \begin{pmatrix} 0 & 0 \\ 1 & 0 \end{pmatrix}, \quad E_{22} = \begin{pmatrix} 0 & 0 \\ 0 & 1 \end{pmatrix}. \tag{12.8}$$

Example 2. Matrix realization of $su(2)$

A real matrix realization of $su(2)$ is

$$E'_{11} = \frac{1}{2}\begin{pmatrix} 1 & 0 \\ 0 & -1 \end{pmatrix}, \quad E_{12} = \begin{pmatrix} 0 & 1 \\ 0 & 0 \end{pmatrix},$$

$$E_{21} = \begin{pmatrix} 0 & 0 \\ 1 & 0 \end{pmatrix}. \tag{12.9}$$

It is of interest in physics to introduce also complex forms of the algebras $u(n)$ and $su(n)$. A common construction is: (i) Retain E_{kk} for $u(n)$ and E'_{kk} for $su(n)$ for the elements with $k = m$. (ii) Take the combinations $(E_{km} + E_{mk})$, $-i(E_{km} - E_{mk})$ for the elements with $k \neq m$. This gives the complex form of $u(n)$

$$
\begin{array}{lll}
E_{kk} & k = 1, \ldots, n & n \\
E_{+,km} = E_{km} + E_{mk} & k < m = 1, \ldots, n & \frac{n(n-1)}{2} \\
E_{-,km} = -i(E_{km} - E_{mk}) & k < m = 1, \ldots, n & \frac{n(n-1)}{2}
\end{array}
\tag{12.10}
$$

The number of elements of each type is written to the far right.

Example 3. Complex form of $u(2)$

The complex form of $u(2)$ is

$$E_{11} = \begin{pmatrix} 1 & 0 \\ 0 & 0 \end{pmatrix}, \quad E_{22} = \begin{pmatrix} 0 & 0 \\ 0 & 1 \end{pmatrix},$$

$$E_{+,12} = \begin{pmatrix} 0 & 1 \\ 1 & 0 \end{pmatrix}, \quad E_{-,12} = \begin{pmatrix} 0 & -i \\ i & 0 \end{pmatrix}. \tag{12.11}$$

Example 4. Complex form of $su(2)$

This algebra is composed of three elements

$$E_{+,12} = E_{12} + E_{21}$$

$$E_{-,12} = -i(E_{12} - E_{21})$$

$$E'_{11} = E_{11} - \frac{1}{2}I. \tag{12.12}$$

The corresponding matrices, denoted by $\sigma_x, \sigma_y, \sigma_z$, are

$$\sigma_x = \begin{pmatrix} 0 & 1 \\ 1 & 0 \end{pmatrix}, \sigma_y = \begin{pmatrix} 0 & -i \\ i & 0 \end{pmatrix}, \sigma_z = \begin{pmatrix} 1 & 0 \\ 0 & -1 \end{pmatrix}. \tag{12.13}$$

The last matrix, E'_{11}, has been multiplied by 2 to conform with the usual normalization of σ_z. They are called Pauli matrices and are widely used in quantum mechanics. The Pauli matrices, together with the unit matrix, I,

$$I = \begin{pmatrix} 1 & 0 \\ 0 & 1 \end{pmatrix}, \tag{12.14}$$

that is the matrices of $u(2)$, form also another algebra, not discussed here, called a Clifford algebra.

Example 5. Complex form of $u(3)$

The complex form of $u(3)$ is

$$E_{+,12} = \begin{pmatrix} 0 & 1 & 0 \\ 1 & 0 & 0 \\ 0 & 0 & 0 \end{pmatrix}, E_{-,12} = \begin{pmatrix} 0 & -i & 0 \\ i & 0 & 0 \\ 0 & 0 & 0 \end{pmatrix}, E_{+,13} = \begin{pmatrix} 0 & 0 & 1 \\ 0 & 0 & 0 \\ 1 & 0 & 0 \end{pmatrix},$$

$$E_{-,13} = \begin{pmatrix} 0 & 0 & -i \\ 0 & 0 & 0 \\ i & 0 & 0 \end{pmatrix}, E_{+,23} = \begin{pmatrix} 0 & 0 & 0 \\ 0 & 0 & 1 \\ 0 & 1 & 0 \end{pmatrix}, E_{-,13} = \begin{pmatrix} 0 & 0 & 0 \\ 0 & 0 & -i \\ 0 & i & 0 \end{pmatrix},$$

$$E_{11} = \begin{pmatrix} 1 & 0 & 0 \\ 0 & 0 & 0 \\ 0 & 0 & 0 \end{pmatrix}, E_{22} = \begin{pmatrix} 0 & 0 & 0 \\ 0 & 1 & 0 \\ 0 & 0 & 0 \end{pmatrix}, E_{33} = \begin{pmatrix} 0 & 0 & 0 \\ 0 & 0 & 0 \\ 0 & 0 & 1 \end{pmatrix}. \tag{12.15}$$

Example 6. Complex form of $su(3)$

In the complex form of $su(3)$ the first six matrices remain the same

$$X_1 = \begin{pmatrix} 0 & 1 & 0 \\ 1 & 0 & 0 \\ 0 & 0 & 0 \end{pmatrix}, X_2 = \begin{pmatrix} 0 & -i & 0 \\ i & 0 & 0 \\ 0 & 0 & 0 \end{pmatrix}, X_3 = \begin{pmatrix} 0 & 0 & 1 \\ 0 & 0 & 0 \\ 1 & 0 & 0 \end{pmatrix},$$

$$X_4 = \begin{pmatrix} 0 & 0 & -i \\ 0 & 0 & 0 \\ i & 0 & 0 \end{pmatrix}, X_5 = \begin{pmatrix} 0 & 0 & 0 \\ 0 & 0 & 1 \\ 0 & 1 & 0 \end{pmatrix}, X_6 = \begin{pmatrix} 0 & 0 & 0 \\ 0 & 0 & -i \\ 0 & i & 0 \end{pmatrix}. \tag{12.16}$$

The last three matrices are replaced by the two matrices

$$X_7 = \frac{1}{3} \begin{pmatrix} 2 & 0 & 0 \\ 0 & -1 & 0 \\ 0 & 0 & -1 \end{pmatrix}, X_8 = \frac{1}{3} \begin{pmatrix} -1 & 0 & 0 \\ 0 & 2 & 0 \\ 0 & 0 & -1 \end{pmatrix}. \tag{12.17}$$

Note that the choice of the traceless matrices is not unique. In applications, often different choices are made. A commonly used choice is

$$X_7 = \begin{pmatrix} 1 & 0 & 0 \\ 0 & -1 & 0 \\ 0 & 0 & 0 \end{pmatrix}, X_8 = \sqrt{\frac{1}{3}} \begin{pmatrix} 1 & 0 & 0 \\ 0 & 1 & 0 \\ 0 & 0 & -2 \end{pmatrix}. \tag{12.18}$$

The corresponding matrices are called Gell-Mann matrices.

12.3 Orthogonal Algebras $so(n)$

A real matrix realization of the algebras $so(n)$ with square $n \times n$ matrices is

$$L_{km} = E_{km} - E_{mk} \quad k < m, \tag{12.19}$$

where the matrices E_{km} are given above.

Example 7. Matrix realization of $so(2)$

A real matrix realization of $so(2)$ is

$$L_{12} = E_{12} - E_{21} = \begin{pmatrix} 0 & 1 \\ -1 & 0 \end{pmatrix}. \tag{12.20}$$

This algebra is Abelian, and can be written in many ways.

Example 8. Matrix realization of $so(3)$

A real matrix realization of $so(3)$ is

$$L_{12} = \begin{pmatrix} 0 & 1 & 0 \\ -1 & 0 & 0 \\ 0 & 0 & 0 \end{pmatrix}, L_{13} = \begin{pmatrix} 0 & 0 & 1 \\ 0 & 0 & 0 \\ -1 & 0 & 0 \end{pmatrix}, L_{23} = \begin{pmatrix} 0 & 0 & 0 \\ 0 & 0 & 1 \\ 0 & -1 & 0 \end{pmatrix}. \tag{12.21}$$

In applications in quantum mechanics, the complex form iL_{km} is used. However, matrix realizations of $so(n)$ are rarely used.

12.4 Symplectic Algebras $sp(2n)$

The construction of symplectic algebras is more involved. It is necessary here to introduce $2n \times 2n$ matrices

$$E_{k,m} = \begin{pmatrix} & & 0 & & \\ & & 0 & & \\ 0\,0\cdots 0 & 1 & 0\,0 \\ & & 0 & & \\ & & \cdots & & \\ & & 0 & & \\ & & 0 & & \end{pmatrix}. \tag{12.22}$$

The rows and columns are labelled by the indices $k, m = 1, \ldots, n; n + 1, \ldots, 2n$.
A real matrix realization of $sp(2n)$ is

$$
\begin{array}{lll}
E_{k,n+k} & k = 1, .., n & n \\
E_{n+k,k} & k = 1, \ldots, n & n \\
E_{k,m} - E_{n+m,n+k} & k, m = 1, \ldots, n & n^2 \\
E_{k,n+m} + E_{m,n+k} & k < m = 1, \ldots, n & \frac{n(n-1)}{2} \\
E_{n+k,m} + E_{n+m,k} & k < m = 1, \ldots, n & \frac{n(n-1)}{2}
\end{array}. \tag{12.23}
$$

The number of elements of each type is shown to the right.

Example 9. The algebra $sp(2)$

In this case $n = 1$. The construction gives

$$E_{12} = \begin{pmatrix} 0 & 1 \\ 0 & 0 \end{pmatrix}; E_{21} = \begin{pmatrix} 0 & 0 \\ 1 & 0 \end{pmatrix}; E_{11} - E_{22} = \begin{pmatrix} 1 & 0 \\ 0 & -1 \end{pmatrix}. \tag{12.24}$$

Note that $sp(2) \sim su(2)$. The matrix realization of $sp(2)$ is identical to that of $su(2)$ given in (12.9), apart from a normalization of the element $E'_{11} = \frac{1}{2}(E_{11} - E_{22})$.

Example 10. The algebra $sp(4)$

In this case $n = 2$. The construction gives the ten elements

$$
\begin{array}{cc}
E_{13}; E_{24}; E_{31}; E_{42} & 4 \\
(E_{11} - E_{33}); (E_{12} - E_{43}); (E_{21} - E_{34}); (E_{22} - E_{44}) & 4. \\
(E_{14} + E_{23}); (E_{32} + E_{41}) & 2
\end{array} \tag{12.25}
$$

where

$$E_{13} = \begin{pmatrix} 0 & 0 & 1 & 0 \\ 0 & 0 & 0 & 0 \\ 0 & 0 & 0 & 0 \\ 0 & 0 & 0 & 0 \end{pmatrix}, etc. \tag{12.26}$$

This provides a real form of $sp(2n)$.

12.5 Basis for the Representation

A basis for the fundamental n dimensional representation of $u(n)$ and $su(n)$ is provided by n-columns vectors

$$\mathcal{B}: \quad \begin{pmatrix} 1 \\ 0 \\ \cdots \\ 0 \\ 0 \end{pmatrix}, \begin{pmatrix} 0 \\ 1 \\ \cdots \\ 0 \\ 0 \end{pmatrix}, \dots, \begin{pmatrix} 0 \\ 0 \\ \cdots \\ 0 \\ 1 \end{pmatrix}. \tag{12.27}$$

Example 11. The basis for $su(2)$

The basis for the two-dimensional representation of $su(2)$ is formed by two columns vectors, often denoted by α, β

$$\alpha = \begin{pmatrix} 1 \\ 0 \end{pmatrix}, \beta = \begin{pmatrix} 0 \\ 1 \end{pmatrix}. \tag{12.28}$$

The matrix realization of the algebra can be re-written in Cartan-Weyl form, introducing the matrices $\sigma_\pm = \frac{1}{2}(\sigma_x \pm i\sigma_y)$,

$$\sigma_+ = \begin{pmatrix} 0 & 1 \\ 0 & 0 \end{pmatrix}, \sigma_- = \begin{pmatrix} 0 & 0 \\ 1 & 0 \end{pmatrix}, \sigma_z = \begin{pmatrix} 1 & 0 \\ 0 & -1 \end{pmatrix}. \tag{12.29}$$

The action of the elements on the basis is

$$\begin{aligned} \sigma_+\alpha = 0 & \quad \sigma_-\alpha = \beta & \quad \sigma_z\alpha = \alpha \\ \sigma_+\beta = \alpha & \quad \sigma_-\beta = 0 & \quad \sigma_z\beta = -\beta \end{aligned}. \tag{12.30}$$

One of the most important applications of $su(2)$ in physics is in the description of particles with spin $S = \frac{1}{2}$. The three spin operators, S_x, S_y, S_z can be simply be written as $S_k = \frac{1}{2}\sigma_k$. The "spin" basis was discussed in Chap. 10 where it was denoted by

$$\left| \begin{array}{cc} su(2) & \supset \quad spin(2) \\ \downarrow & \quad \downarrow \\ S = \frac{1}{2} & \quad M_S = \pm\frac{1}{2} \end{array} \right). \tag{12.31}$$

The relation between the basis here and that discussed in Chap. 10 is

$$\alpha \equiv \left| \frac{1}{2}, +\frac{1}{2} \right\rangle, \beta \equiv \left| \frac{1}{2}, -\frac{1}{2} \right\rangle. \tag{12.32}$$

Example 12. The basis for su(3)

The basis for the fundamental representation of $su(3)$ is formed by three column vectors

$$\alpha_1 = \begin{pmatrix} 1 \\ 0 \\ 0 \end{pmatrix}, \alpha_2 = \begin{pmatrix} 0 \\ 1 \\ 0 \end{pmatrix}, \alpha_3 = \begin{pmatrix} 0 \\ 0 \\ 1 \end{pmatrix}. \tag{12.33}$$

The matrix realization of $su(3)$ was given in Example 6. In applications to particle physics, the Gell-Mann matrices were used. This basis was discussed in Chap. 10, where it was denoted by

$$\begin{vmatrix} su(3) & \supset & su(2) & \supset & spin(2) \\ \downarrow & & \downarrow & & \downarrow \\ [\lambda_1, \lambda_2] \equiv [1] & & T = 0, \frac{1}{2} & & M_T = 0, \pm\frac{1}{2} \end{vmatrix}. \tag{12.34}$$

The relation between the basis (12.33) and that discussed in Chap. 10 is

$$\alpha_1 = \left|[1], \frac{1}{2}, +\frac{1}{2}\right\rangle \equiv u, \alpha_2 = \left|[1], \frac{1}{2}, -\frac{1}{2}\right\rangle \equiv d, \alpha_3 = |[1], 0, 0\rangle \equiv s, \tag{12.35}$$

where the notation u, d, s used in particle physics has also been included.

For representations of $u(n)$ and $su(n)$ other than the fundamental representation, matrix realizations are seldom used.

12.6 Casimir Operators

The Casimir operators can be constructed from matrices. For the algebras $u(n)$ and $su(n)$, the unit matrix I commutes with all elements and thus is a Casimir operator.

Example 13. The Casimir operator of su(2)

For applications to physics it is of particular interest to construct the Casimir operator of $su(2)$. This algebra has only one invariant, $C_2(su(2))$. It can be easily shown that the invariant is

$$C_2(su(2)) = \sigma_x^2 + \sigma_y^2 + \sigma_z^2 = 3I. \tag{12.36}$$

(The condition $\sigma_i^2 = I$ is one of the axioms defining a Clifford algebra.) The eigenvalues of the Casimir operator in the basis are

$$\langle\alpha| \sum_k \sigma_k^2 |\alpha\rangle = 3, \quad \langle\beta| \sum_k \sigma_k^2 |\beta\rangle = 3. \tag{12.37}$$

In application in physics, the algebra $su(2)$ often describes spin $\frac{1}{2}$ particles. The eigenvalues of the Casimir invariant in the two-dimensional representation of $su(2)$ is

$$\langle C_2(su(2)) \rangle = \frac{3}{4}. \tag{12.38}$$

This value agrees with the eigenvalues of the Casimir operator in the generic representation $|J, M\rangle$ of $su(2)$ given in Chap. 10,

$$\langle C_2(su(2)) \rangle = J(J + 1), \tag{12.39}$$

since here $J = S = \frac{1}{2}$.

Chapter 13
Coset Spaces

Coset spaces introduced in Chap. 5 are used in many applications in physics. Among
other things, they provide a "geometry" to the algebraic models discussed in Chap. 9
by associating geometric variables to the coset decomposition (5.1) of the algebra
g into $h \oplus p$. Denoting by $|\Lambda\rangle$ the irreducible representations of g, one chooses
among them a state, called the extremal state $|\Lambda_{ext}\rangle$, such that it is annihilated by
the largest number of vectors in the algebra. The "geometric' coset variables η_i are
defined by

$$|\eta_i\rangle = \exp\left[\sum_i \eta_i p_i\right] |\Lambda_{ext}\rangle, \qquad p_i \in p. \tag{13.1}$$

In the following sections, an explicit construction of the coset spaces
$U(n)/U(n-1) \otimes U(1)$, $SO(n)/SO(n-1)$ and $SO(n+2)/SO(n) \otimes SO(2)$ is given.
These spaces are denoted either by the groups $G/G' \otimes G''$ or by the corresponding
algebras $g/g' \oplus g''$.

13.1 Coset Spaces $U(n)/U(n-1) \otimes U(1)$

The simplest way to construct these spaces is within a bosonic realization. Let g be
$u(n)$ and consider the realization

$$g = u(n); \quad g \doteq G_{\alpha\beta} = b_\alpha^\dagger b_\beta; \quad \alpha, \beta = 1, \ldots, n. \tag{13.2}$$

Let the stability algebra h be

$$h = u(n-1) \oplus u(1). \tag{13.3}$$

© Springer-Verlag Berlin Heidelberg 2015
F. Iachello, *Lie Algebras and Applications*, Lecture Notes in Physics 891,
DOI 10.1007/978-3-662-44494-8_13

This stability algebra, called the maximal stability algebra, can be simply constructed by selecting one of the bosons, say 1, and writing

$$h \doteq b_1^\dagger b_1, b_\alpha^\dagger b_\beta, \qquad \alpha, \beta = 2, \ldots, n. \tag{13.4}$$

The remainder p in the Cartan decomposition $g = h \oplus p$ is then

$$p \doteq b_1^\dagger b_\alpha, b_\alpha^\dagger b_1, \qquad \alpha = 2, \ldots, n. \tag{13.5}$$

The properties of h and p are

$$[h, h] \subset h, \quad [h, p] \subset p, \quad [p, p] \subset h. \tag{13.6}$$

The irreducible representations of $u(n)$ that can be constructed with boson operators are the totally symmetric representations

$$|N\rangle \equiv \left| \underbrace{N, 0, 0, \ldots, 0}_{n} \right\rangle. \tag{13.7}$$

The extremal state is

$$|\Lambda_{ext}\rangle = \frac{1}{\sqrt{N!}} \left(b_1^\dagger \right)^N |0\rangle. \tag{13.8}$$

13.2 Coherent (or Intrinsic) States

Coset spaces can be constructed by introducing the states (13.1), called coherent or intrinsic states. Several types of coherent states have been used in applications.

13.2.1 'Algebraic' Coherent States

We introduce $n - 1$ complex variables, $\eta_\alpha, \alpha = 2, .., n$. Algebraic coherent states are defined as

$$|N; \eta_\alpha\rangle = \left[\exp \left(\eta_\alpha b_\alpha^\dagger b_1 - \eta_\alpha^* b_1^\dagger b_\alpha \right) \right] \frac{1}{\sqrt{N!}} \left(b_1^\dagger \right)^N |0\rangle. \tag{13.9}$$

All states of the irrep $[N]$ can be generated in this way. They are called 'algebraic' because the variables η_α parametrize the elements p_i of the algebra.

13.2.2 'Group' Coherent States

We introduce $n - 1$ complex variables $\zeta_\alpha, \alpha = 2, \ldots, n$. Group coherent states are defined as

$$|N; \zeta_\alpha\rangle = \frac{1}{\sqrt{N!}} \left[(1 - \zeta_\alpha^* \zeta_\alpha)^{1/2} b_1^\dagger + \zeta_\alpha b_\alpha^\dagger \right]^N |0\rangle . \qquad (13.10)$$

These states are called 'group' coherent states because ζ_α is the matrix element of the group element that transforms $b_1^\dagger |0\rangle$ into $b_\alpha^\dagger |0\rangle$.

Introducing the redundant variable ζ_1, one has

$$|\zeta_1|^2 + |\zeta_2|^2 + \ldots + |\zeta_n|^2 = 1. \qquad (13.11)$$

The space ζ is therefore compact

$$|\zeta|^2 = \zeta_\alpha^* \zeta_\alpha \leq 1, \quad \alpha = 2, \ldots, n. \qquad (13.12)$$

A relation between the group and the algebraic variables can be found by expanding the exponential in (13.9) and noting that the extremal state (13.8) is annihilated by all boson operators $b_\alpha, \alpha = 2, \ldots, n$,

$$b_\alpha |\Lambda_{ext}\rangle = 0, \quad \alpha = 2, .., n. \qquad (13.13)$$

The state (13.9) can then be rewritten as

$$|N; \eta_\alpha\rangle = \frac{1}{\sqrt{N!}} \left[(\cos |\eta|) b_1^\dagger + \frac{\sin |\eta|}{|\eta|} \eta_\alpha b_\alpha^\dagger \right]^N |0\rangle , \quad |\eta| = (\eta_\alpha^* \eta_\alpha)^{1/2} , \qquad (13.14)$$

yielding

$$\zeta_\alpha = \eta_\alpha \frac{\sin |\eta|}{|\eta|}. \qquad (13.15)$$

The group coherent states have the interesting property that the resolution of the identity can be written as

$$1 = D(N) \frac{(n-1)!}{(2\pi i)^{n-1}} \int d^{n-1}\zeta d^{n-1}\zeta^* |N; \zeta\rangle \langle N; \zeta| , \qquad (13.16)$$

where $D(N)$ is the dimension of the representation $[N]$ discussed in Chap. 6. The resolution of the identity for the 'algebraic' and 'projective' states is similar to (13.16) but with a different integration measure.

13.2.3 'Projective' Coherent States

We introduce here $n - 1$ complex variables $\vartheta_\alpha, \alpha = 2, \ldots, n$. The projective coherent states are

$$
\begin{aligned}
|N; \vartheta_\alpha\rangle &= \exp\left[\vartheta_\alpha b_\alpha^\dagger b_1\right] \frac{1}{\sqrt{N!}} \left(b_1^\dagger\right)^N |0\rangle \\
&= \frac{1}{\sqrt{N!}} \left[b_1^\dagger + \vartheta_\alpha b_\alpha^\dagger\right]^N |0\rangle .
\end{aligned} \tag{13.17}
$$

The relation between projective variables ϑ_α and group variables ζ_α is

$$
\vartheta_\alpha = \zeta_\alpha \left(1 - \zeta_\alpha^* \zeta_\alpha\right)^{-1/2} . \tag{13.18}
$$

The states are called 'projective' because of the projective transformation

$$
|\vartheta_\alpha|^2 = \frac{|\zeta_\alpha|^2}{1 - |\zeta_\alpha|^2} . \tag{13.19}
$$

Projective coherent states are often used in applications. Normalized projective coherent states, also called intrinsic states, are defined as

$$
|N; \vartheta_\alpha\rangle = \frac{1}{\sqrt{N!}} \left[\frac{1}{\left(1 + |\vartheta|^2\right)^{1/2}} \left(b_1^\dagger + \vartheta_\alpha b_\alpha^\dagger\right)\right]^N |0\rangle . \tag{13.20}
$$

The states (13.20) are also called 'condensate' coherent states since they are obtained by applying to the vacuum $|0\rangle$ the 'condensate' creation operator

$$
b_c^\dagger = \frac{1}{\left(1 + |\vartheta|^2\right)^{1/2}} \left(b_1^\dagger + \vartheta_\alpha b_\alpha^\dagger\right) \tag{13.21}
$$

N times,

$$
|N; \vartheta_\alpha\rangle = \frac{1}{\sqrt{N!}} \left(b_c^\dagger\right)^N |0\rangle . \tag{13.22}
$$

13.2.4 Coset Spaces

The coherent states of the previous subsection provide parametrizations of the coset spaces $U(n)/U(n-1) \otimes U(1)$ in terms of the $n-1$ complex variables η_α (algebraic), ζ_α (group), ϑ_α (projective). The dimension of these spaces is $2(n-1)$ as in Table 5.1,

$$\dim[U(n)/U(n-1) \otimes U(1)] = 2(n-1). \qquad (13.23)$$

Particularly important for applications are: (i) the complex projective spaces PC_{n-1} with $n-1$ complex variables ϑ_α and (ii) the group spaces, ζ_α. Introducing the redundant variable ζ_1 as in (13.11), one can convert these spaces into the complex spaces C_n with constraint

$$\zeta_1 \zeta_1^* + \ldots + \zeta_n \zeta_n^* = 1. \qquad (13.24)$$

In some applications, it is also of interest to construct the coset spaces $U(n-1,1)/U(n-1) \otimes U(1)$. These have the same dimension as (13.23)

$$\dim[U(n-1,1)/U(n-1) \otimes U(1)] = 2(n-1) \qquad (13.25)$$

and can be converted into complex spaces C_n as before, but with the constraint

$$\zeta_1 \zeta_1^* - \zeta_2 \zeta_2^* - \ldots - \zeta_n \zeta_n^* = 1. \qquad (13.26)$$

13.3 Coherent States of $u(n), n = Even$

In many applications it is of interest to construct coherent states for rotationally invariant problems. In these cases, it is convenient to introduce the Racah form of the Lie algebra, discussed in Chap. 9. The algebra $u(n), n =$even, can be constructed in Racah form by introducing a scalar boson operator, s, and a boson operator, b_l, that transforms as the representation $|l\rangle$ of $so(3)$. Coherent states can be simply constructed in this case by selecting the s boson as generating $u(1)$ and writing

$$h = u(2l+1) \oplus u(1). \qquad (13.27)$$

The $u(1)$ algebra in (13.27) is simply composed of $s^\dagger s$, while $u(2l+1)$ is the boson realization $b_{l,m}^\dagger b_{l,m'}$ of Chap. 9. Particularly interesting cases for applications are those with $l = 0$ $(u(2))$, $l = 1$ $(u(4))$, and $l = 2$ $(u(6))$.

13.3.1   Coherent States of $u(2)$

We introduce a boson realization of $u(2)$ in terms of two (scalar) bosons σ, τ. This realization is the same as in (9.18), except for renaming $b_1 \to \sigma$, $b_2 \to \tau$. The stability algebra and the remainder are simply constructed

$$g = u(2) \doteq \sigma^\dagger \tau, \tau^\dagger \sigma, \tau^\dagger \tau, \sigma^\dagger \sigma$$
$$h = u(1) \oplus u(1) \doteq \tau^\dagger \tau, \sigma^\dagger \sigma$$
$$p \doteq \sigma^\dagger \tau, \tau^\dagger \sigma. \tag{13.28}$$

In this case we have one complex variable ζ and the group states are

$$|N; \zeta\rangle = \frac{1}{\sqrt{N!}} \left[(1 - \zeta^* \zeta)^{1/2} \sigma^\dagger + \zeta \tau^\dagger \right]^N |0\rangle. \tag{13.29}$$

The complex variable ζ can be split into a coordinate q and a momentum p

$$\zeta = \frac{1}{\sqrt{2}} (q + ip), \quad \zeta^* = \frac{1}{\sqrt{2}} (q - ip), \tag{13.30}$$

describing motion in one dimension.

The normalized 'projective' state, also called intrinsic state, is written as

$$|N; r\rangle = \frac{1}{\sqrt{N!}} \left[\frac{1}{\sqrt{1 + r^2}} (\sigma^\dagger + r \tau^\dagger) \right]^N |0\rangle, \tag{13.31}$$

where the variable $r \equiv \vartheta$ is called the intrinsic variable.

13.3.2   Coherent States of $u(4)$

We construct the algebra of $u(4)$ in Racah form by introducing a scalar boson, σ, and a vector boson, $\pi_\mu, \mu = 0, \pm 1$, that transforms as the representation $l = 1$ of $so(3)$. The vector boson is denoted here by π_μ instead of p_μ of (9.112), not to confuse it with the momentum. The elements of the Lie algebra $g \equiv u(4)$ can be split into a stability algebra $h = u(3) \oplus u(1)$ and a remainder p as

$$\left.\begin{array}{l}\left.\begin{array}{l}\left(\sigma^\dagger \times \tilde{\sigma}\right)_0^{(0)} \Big\} u(1) \\[1.2em] \left(\pi^\dagger \times \tilde{\pi}\right)_0^{(0)} \\[1.2em] \left(\pi^\dagger \times \tilde{\pi}\right)_\kappa^{(1)} \Big\} u(3) \\[1.2em] \left(\pi^\dagger \times \tilde{\pi}\right)_\kappa^{(2)} \end{array}\right\} h \end{array}\right.$$

$$\left.\begin{array}{l}\left(\pi^\dagger \times \tilde{\sigma}\right)_\kappa^{(1)} \\[1.2em] \left(\sigma^\dagger \times \tilde{\pi}\right)_\kappa^{(1)} \end{array}\right\} p. \tag{13.32}$$

The phase convention in (13.32) is $\tilde{\pi}_\mu = (-)^\mu \pi_{-\mu}, \tilde{\sigma} = \sigma$. Also, instead of the operators in (13.32), one can use as a remainder p,

$$\left(\pi^\dagger \times \tilde{\sigma} + \sigma^\dagger \times \tilde{\pi}\right)$$
$$i \left(\pi^\dagger \times \tilde{\sigma} - \sigma^\dagger \times \tilde{\pi}\right). \tag{13.33}$$

There are in this case 3 complex variables $\zeta_\mu, \mu = 0, \pm1$. The group states are

$$|N; \zeta_\mu\rangle = \frac{1}{\sqrt{N!}} \left[\left(1 - |\zeta|^2\right)^{1/2} \sigma^\dagger + \left(\zeta \cdot \pi^\dagger\right)\right]^N |0\rangle, \tag{13.34}$$

where the dot denotes scalar products with respect to $so(3)$

$$A^{(k)} \cdot B^{(k)} = \sum_\mu (-)^\mu A_\mu^{(k)} B_{-\mu}^{(k)}. \tag{13.35}$$

The complex variables ζ_μ can be rewritten in terms of coordinates and momenta in a three-dimensional space as

$$\zeta_\mu = \frac{1}{\sqrt{2}} \left(q_\mu + ip_\mu\right), \tag{13.36}$$

with phases such that

$$\zeta_\mu^* = \frac{1}{\sqrt{2}} \left(q_\mu^* - ip_\mu^*\right), \quad q_\mu^* = (-)^\mu q_{-\mu}, \quad p_\mu^* = (-)^\mu p_{-\mu}. \tag{13.37}$$

The normalized coherent state is

$$|N;\vartheta_\mu\rangle = \frac{1}{\sqrt{N!}}\left[\frac{1}{\left(1+|\vartheta|^2\right)^{1/2}}\left(\sigma^\dagger + \vartheta_\mu \pi_\mu^\dagger\right)\right]^N |0\rangle. \tag{13.38}$$

Instead of using the three variables ϑ_μ, $\mu = 0, \pm1$, one can use the two angles (θ, ϕ) defining the orientation of the vector ϑ, and an intrinsic variable r. The intrinsic state can then be written in the form

$$|N;r\rangle = \frac{1}{\sqrt{N!}}\left[\frac{1}{\sqrt{1+r^2}}\left(\sigma^\dagger + r\pi_0^\dagger\right)\right]^N |0\rangle, \tag{13.39}$$

containing only the intrinsic variable r.

13.3.3   Coherent States of $u(6)$

We construct the algebra $u(6)$ in Racah form by introducing a scalar boson, s, and a quadrupole boson d_μ, $\mu = 0, \pm1, \pm2$, that transforms as $l = 2$ of $so(3)$. The elements of the Lie algebra $g \equiv u(6)$ can be split into a stability algebra h and a remainder p as

$$\left.\begin{array}{c}
\left(s^\dagger \times \tilde{s}\right)_0^{(0)}\Big\}\, u(1) \\[2ex]
\left(d^\dagger \times \tilde{d}\right)_0^{(0)} \\[2ex]
\left(d^\dagger \times \tilde{d}\right)_\kappa^{(1)} \\[2ex]
\left(d^\dagger \times \tilde{d}\right)_\kappa^{(2)}\,\Big\}\, u(5) \\[2ex]
\left(d^\dagger \times \tilde{d}\right)_\kappa^{(3)} \\[2ex]
\left(d^\dagger \times \tilde{d}\right)_\kappa^{(4)}
\end{array}\right\}\, h$$

$$\left.\begin{array}{c}
\left(d^\dagger \times \tilde{s}\right)_\kappa^{(2)} \\[2ex]
\left(s^\dagger \times \tilde{d}\right)_\kappa^{(2)}
\end{array}\right\}\, p, \tag{13.40}$$

with the phase convention $\tilde{d}_\mu = (-)^\mu d_{-\mu}$, $\tilde{s} = s$. Also here, instead of the remainder in (13.40), one can use

$$\left(d^\dagger \times \tilde{s} + s^\dagger \times \tilde{d}\right)^{(2)}_\kappa$$

$$i\left(d^\dagger \times \tilde{s} - s^\dagger \times \tilde{d}\right)^{(2)}_\kappa . \tag{13.41}$$

The group coherent states are written in terms of 5 complex variables ζ_μ, $\mu = 0, \pm1, \pm2$ as

$$|N; \zeta_\mu\rangle = \frac{1}{\sqrt{N!}}\left[\left(1 - |\zeta|^2\right)^{1/2} s^\dagger + \left(\zeta \cdot d^\dagger\right)\right]^N |0\rangle . \tag{13.42}$$

Introducing coordinates and momenta

$$\zeta_\mu = \frac{1}{\sqrt{2}}\left(q_\mu + ip_\mu\right)$$

$$\zeta_\mu^* = \frac{1}{\sqrt{2}}\left(q_\mu - ip_\mu\right), \tag{13.43}$$

one can study motion in a five-dimensional space.

The normalized intrinsic state is

$$|N; \vartheta_\mu\rangle = \frac{1}{\sqrt{N!}}\left[\frac{1}{\left(1 + |\vartheta|^2\right)^{1/2}}\left(s^\dagger + \vartheta_\mu d_\mu^\dagger\right)\right]^N |0\rangle . \tag{13.44}$$

Instead of using the five variables ϑ_μ, $\mu = 0, \pm1, \pm2$, one can use the three Euler angles $(\theta_1, \theta_2, \theta_3)$ defining the orientation of the quadrupole tensor ϑ in space, and two intrinsic variables, β and γ, called Bohr variables. The intrinsic state can then be written in the form

$$|N; \beta, \gamma\rangle = \frac{1}{\sqrt{N!}}\left[\frac{1}{\sqrt{1 + \beta^2}}\left[s^\dagger + \beta\left(\cos\gamma d_0^\dagger + \frac{1}{\sqrt{2}}\sin\gamma(d_{+2}^\dagger + d_{-2}^\dagger)\right)\right]\right]^N$$
$$\times |0\rangle , \tag{13.45}$$

containing only the intrinsic variables β and γ. Details of the definition of intrinsic Bohr variables are given in Iachello and Arima (1987).

13.4 Coherent States of $u(n), n = Odd$

Coherent states of $u(n), n = $ odd, can be constructed in a similar fashion as those of
$n = $ even, except that now one divides the boson operators into a scalar boson s and
boson operators b_l with an even number of components. These boson operators
are no longer tensors under $so(3)$, but rather tensors under the algebra of two-
dimensional rotations, $so(2)$, and are called cylindrical or circular bosons. Important
cases are $l = 1$, doublet boson with components $\mu = \pm 1$ under $so(2)$, and $l = 2$,
quadruplet boson with components $\mu = \pm 2, \pm 1$ under $so(2)$, by means of which
one can construct bosonic coherent states of $u(3)$ and $u(5)$.

13.4.1 Coherent States of $u(3)$

We introduce here a singlet boson operator σ and a doublet τ_x, τ_y. The doublet can
be rewritten as in (9.100). The cylindrical boson operators

$$\tau^\dagger_\pm = \mp \frac{1}{\sqrt{2}} \left(\tau^\dagger_x \pm i \tau^\dagger_y \right), \tag{13.46}$$

transform as the representations $\mu = \pm 1$ of $so(2)$. The algebra $u(3)$ and its maximal
stability algebra $u(2) \oplus u(1)$ are, in the notation of Sect. 9.5.6,

$$g \doteq \hat{n}_s, \hat{n}, \hat{l}, \hat{Q}_+, \hat{Q}_-, \hat{D}_+, \hat{D}_-, \hat{R}_+, \hat{R}_-$$
$$h \doteq \hat{n}_s; \hat{n}, \hat{l}, \hat{Q}_+, \hat{Q}_-$$
$$p \doteq \hat{D}_+, \hat{D}_-, \hat{R}_+, \hat{R}_-. \tag{13.47}$$

Introducing the two-dimensional complex variable $\zeta_\mu, \mu = \pm 1$, one can write the
group states as

$$|N; \zeta_\mu\rangle = \frac{1}{\sqrt{N!}} \left[\left(1 - |\zeta|^2 \right)^{1/2} \sigma^\dagger + \left(\zeta \cdot \tau^\dagger \right) \right]^N |0\rangle, \tag{13.48}$$

where the dot represents two-dimensional scalar product, $A \cdot B = A_x B_x + A_y B_y$.
    The intrinsic state has also in this case the structure of (13.20) with $\vartheta_\mu, \mu = \pm 1$.
However, when going to intrinsic coordinates, it is convenient to use cylindrical
coordinates, $x = r \cos \theta, y = r \sin \theta$, and define the frame in such a way that
$\theta = 0$. The intrinsic state is then

$$|N; r\rangle = \frac{1}{\sqrt{N!}} \left[\frac{1}{\sqrt{1+r^2}} \left(\sigma^\dagger + r\tau^\dagger_x \right) \right]^N |0\rangle. \tag{13.49}$$

13.5   Generalized Coherent States

The coherent states described in the previous sections represent only a portion of all coherent states. In physical applications, they describe the ground state of a bosonic system. Coherent states describing multiply excited states of bosonic systems have been constructed. A discussion of the generalized coherent states of $u(3)$ constructed with a scalar boson, σ, and a doublet boson, $\tau_\pm$, is given in Caprio (2005). A general theory of coherent states is given in Perelomov (1986). A review is given in Zhang et al. (1990).

13.6   The Geometry of Algebraic Models

The coherent states introduced in the previous sections form a basis for the study of the geometry of algebraic models discussed in Chap. 9. In these models, the Hamiltonian and other operators are expanded into elements X_i of a Lie algebra g, $X_i \in g$, usually taken as $u(n)$,

$$H = E_0 + \sum_i \varepsilon_i X_i + \sum_{ij} v_{ij} X_i X_j + \ldots$$

$$T = T_0 + \sum_i t_i X_i + \ldots . \tag{13.50}$$

The coherent states for these models are the coset states $u(n)/u(n-1) \oplus u(1)$. They have $n-1$ complex variables $\zeta \equiv (\zeta_1, \zeta_2, \ldots, \zeta_{n-1})$. The expectation value of the Hamiltonian and other operators in the coherent state is called the 'classical' limit of the algebraic Hamiltonian

$$H_{cl}\left(\zeta, \zeta^*\right) = \frac{\langle [N]; \zeta \,|H|\, [N]; \zeta \rangle}{\langle [N]; \zeta | [N]; \zeta \rangle}. \tag{13.51}$$

By introducing coordinates and momenta, as in (13.30), $\zeta = \frac{1}{\sqrt{2}}(q + ip)$, one can construct a Hamiltonian system with H a function of p and q, $H(p, q)$. This Hamiltonian is not in the canonical Schrödinger form

$$H(p, q) = \frac{p^2}{2m} + V(q), \tag{13.52}$$

with separation into a kinetic term, independent of q, and a potential term, independent of p. Nonetheless, it is of great interest in applications, especially in the study of phase transitions in algebraic models. A review of these applications is given in Cejnar and Iachello (2007).

13.7 Coset Spaces $SO(n + m)/SO(n) \otimes SO(m)$

These coset spaces are particularly important in the description of space coordinates in R_n and also for the construction of real projective spaces, PR_n. The dimension of the real coset $SO(n + m)/SO(n) \otimes SO(m)$ is

$$\dim [SO(n + m)/SO(n) \otimes SO(m)] = \frac{1}{2} (n + m) (n + m - 1)$$

$$-\frac{1}{2}n (n - 1) - \frac{1}{2}m (m - 1) = nm. \tag{13.53}$$

The spaces $SO(n + 1)/SO(n)$ have dimension n and are used in the description of problems with rotational invariance both in classical and quantum mechanics.

13.7.1 The Coset Space $SO(3)/SO(2)$: The Sphere S^2

We consider here the coset space $SO(3)/SO(2)$ with dimension 2. This space can also be mapped onto the coset space $U(2)/U(1) \otimes U(1)$ and $SU(2)/U(1)$. The elements of the Lie algebra $so(3)$ can be written as in (2.10). The stability algebra in the decomposition $g = h \oplus p$ is formed by the single element J_z and the remainder p is J_+, J_-,

$$g = so(3) \doteq J_+, J_-, J_z$$
$$h = so(2) \doteq J_z$$
$$p \doteq J_+, J_- \tag{13.54}$$

We denote the representations of $so(3)$ by $|j\rangle$ with components $|j, m\rangle$. We take as extremal state

$$|\Lambda_{ext}\rangle \equiv |j, -j\rangle . \tag{13.55}$$

The coherent states of $so(3)/so(2) \approx u(2)/u(1) \oplus u(1) \approx su(2)/u(1)$ can be written in various ways. The 'algebraic' coherent states of $u(2)/u(1) \oplus u(1)$ can be written as in (13.9) in terms of the complex variable η

$$|j; \eta, \eta^*\rangle = \exp [\eta J_+ - \eta^* J_-] |j, -j\rangle . \tag{13.56}$$

By introducing the real variables θ, ϕ through

$$\eta = e^{-i\phi} \frac{\theta}{2} , \quad \eta^* = e^{i\phi} \frac{\theta}{2}, \tag{13.57}$$

they can be rewritten as

$$|j; \theta, \phi\rangle = \exp\left[e^{-i\phi}\frac{\theta}{2}J_+ - e^{i\phi}\frac{\theta}{2}J_-\right]|j, -j\rangle. \qquad (13.58)$$

These states, called Bloch states, are used in quantum mechanics. Their overlap is

$$\langle j; \theta', \phi'|j; \theta, \phi\rangle = \left[\cos\frac{\theta'}{2}\cos\frac{\theta}{2} + e^{-i(\phi-\phi')}\sin\frac{\theta'}{2}\sin\frac{\theta}{2}\right]^{2j}. \qquad (13.59)$$

Instead of the Cartan form of the algebra J_+, J_-, J_z, one can use the Cartesian form J_x, J_y, J_z with $J_\pm = J_x \pm iJ_y$. The algebraic coherent states can then be written as

$$|j; \eta, \eta^*\rangle = \exp\left[(\eta - \eta^*)J_x + i(\eta + \eta^*)J_y\right]|j, -j\rangle. \qquad (13.60)$$

By introducing the real and complex part of η, $\eta = \alpha + i\beta$, $\eta^* = \alpha - i\beta$, one can rewrite (13.60) as

$$|j; \beta_1, \beta_2\rangle = \exp\left[\beta_1 X_1 + \beta_2 X_2\right]|j, -j\rangle, \qquad (13.61)$$

where $X_1 = iJ_x$, $X_2 = iJ_y$ are elements of the real Lie algebra $so(3)$, (1.9), and $\beta_1 = 2i\beta$, $\beta_2 = 2i\alpha$. We see here explicitly how all the states of the representation $|j\rangle$ can be generated starting from the extremal state $|j, -j\rangle$, Fig. 13.1.

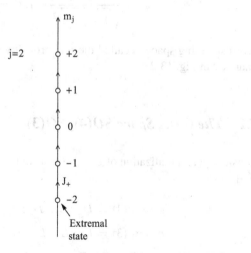

Fig. 13.1 States of the representation $j = 2$ generated by the action of the remainder p on the extremal state $|2, -2\rangle$

Parametrizations of the Coset Space $SO(3)/SO(2)$

The 'algebraic' parametrization of the coset space $SO(3)/SO(2)$ is in terms of two parameters, β_1, β_2. This space is just R_2.

The 'group' parametrization of the coset space $SO(3)/SO(2)$ can be obtained from β_1, β_2 with the change of variables

$$x_i = \beta_i \frac{\sin\left(\sum_{i'} \beta_{i'}^2\right)^{1/2}}{\left(\sum_{i'} \beta_{i'}^2\right)^{1/2}} \quad , \quad i, i' = 1, 2. \tag{13.62}$$

Introducing the redundant variable x_3

$$x_3 = \pm\sqrt{1 - x_1^2 - x_2^2}, \tag{13.63}$$

one has

$$x_1^2 + x_2^2 + x_3^2 = 1. \tag{13.64}$$

This space is thus the two-sphere embedded in R_3, $S^2 \subset R_3$. Points on the unit sphere can be parametrized by two angles θ, ϕ through

$$x_1 = \sin\theta\cos\phi \,, \quad x_2 = \sin\theta\sin\phi \,, \quad x_3 = \cos\theta. \tag{13.65}$$

The 'projective' parametrization is obtained by the change of variables

$$z_i = \frac{x_i}{\sqrt{1 - \sum_{i'} x_{i'}^2}}, \quad i, i' = 1, 2. \tag{13.66}$$

The corresponding space is called the real projective space, PR_2. The situation is summarized in Fig. 13.2.

13.7.2   The Coset Space $SO(4)/SO(3)$

We consider here a realization of $g = so(4)$ as in (1.29). The algebra g can be split into $h \oplus p$ as

$$g = so(4) \doteq D_+, D_-, D_z \,; \, L_+, L_-, L_z$$
$$h = so(3) \doteq L_+, L_-, L_z$$
$$p \doteq D_+, D_-, D_z, \tag{13.67}$$

with $[L, L] \subset L$, $[D, L] \subset D$, $[D, D] \subset L$. The representations of $so(4)$ are labelled as in (6.59)

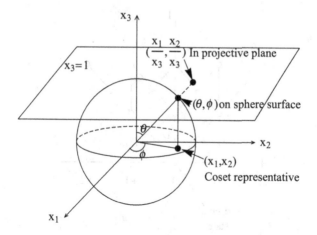

Fig. 13.2 Group variables x_1, x_2, x_3 with $x_1^2 + x_2^2 + x_3^2 = 1$ or (θ, ϕ) for the sphere $S^2 \subset R_3$, and projective variables $\frac{x_1}{x_3}, \frac{x_2}{x_3}$ for PR_2

$$\left| \begin{array}{ccc} so(4) & \supset so(3) & \supset so(2) \\ \downarrow & \downarrow & \downarrow \\ (\omega_1, \omega_2) & L & M_L \end{array} \right). \tag{13.68}$$

We consider the symmetric representation $(\omega, 0)$ with branching $L = 0, 1, .., \omega$ and $-L \le M_L \le +L$ and take as extremal state $L = 0$, $M_L = 0$. The algebraic coherent states for the representation $(\omega, 0)$ are

$$|(\omega, 0); \eta_+, \eta_-, \eta_z\rangle \equiv |(\omega, 0); \boldsymbol{\eta}\rangle$$

$$= \exp\left[\eta_+ D_+ + \eta_- D_- + \eta_z D_z\right] |(\omega, 0); 0, 0\rangle$$

$$= \exp\left[\boldsymbol{\eta} \cdot \mathbf{D}\right] |(\omega, 0); 0, 0\rangle. \tag{13.69}$$

All states can be obtained from (13.69) as shown schematically in Fig. 13.3.

Parametrizations of the Coset Space $SO(4)/SO(3)$

The algebraic parametrization of the coset space $SO(4)/SO(3)$ is in terms of three real variables $\beta_1, \beta_2, \beta_3$. This space is just R_3.

The group parametrization is in terms of variables

$$x_i = \beta_i \frac{\sin\left(\sum_{i'} \beta_{i'}^2\right)^{1/2}}{\left(\sum_{i'} \beta_{i'}^2\right)^{1/2}}, \quad i, i' = 1, 2, 3. \tag{13.70}$$

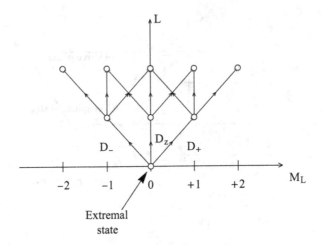

Fig. 13.3 The states of the representation $(2, 0)$ of $so(4)$ obtained from the extremal state $L = 0$, $M_L = 0$

Adding a redundant variable $x_4 = \pm\sqrt{1 - x_1^2 - x_2^2 - x_3^2}$, we see that the group space is the three-sphere, $S^3 \subset R_4$. One can also use three angles θ, ϕ, ψ, called Pauli angles, to parametrize this space with states $|(\omega, 0); \theta, \phi, \psi\rangle$.

The projective parametrization is in terms of the projective variables

$$z_i = \frac{x_i}{\sqrt{1 - \sum_{i'} x_{i'}^2}}, \qquad i, i' = 1, 2, 3. \tag{13.71}$$

This space is the projective real space PR_3.

13.7.3 The Coset Spaces $SO(n+2)/SO(n) \otimes SO(2)$

The coset spaces $SO(n + 1)/SO(n)$ of the preceding Sect. 13.7.2 are those most used in quantum mechanics and are simply constructed in terms of n real variables. As an example of more elaborate coset spaces, we consider here the spaces $SO(n + 2)/SO(n) \otimes SO(2)$ with $2n$ real variables. In particular, we explicitly construct the space $SO(4)/SO(2) \otimes SO(2)$.

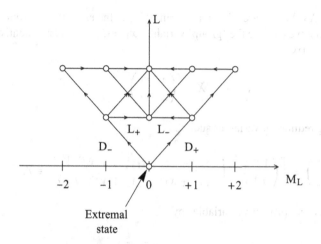

Fig. 13.4 The states of the representation $(2,0)$ obtained from the extremal state $L = 0, M_L = 0$ by acting with the remainder p of (13.72)

The Coset Space $SO(4)/SO(2) \otimes SO(2)$

The algebra $so(4)$ composed of $D_\pm, D_z, L_\pm, L_z$ of the previous subsection is decomposed here into $g = h \oplus p$, with $h = so(2) \oplus so(2)$, as

$$g = so(4) \doteq D_+, D_-, D_z ; L_+, L_-, L_z$$
$$h = so(2) \oplus so(2) \doteq D_z ; L_z$$
$$p \doteq D_+, D_- ; L_+, L_- \tag{13.72}$$

The algebraic coherent states are

$$|(\omega,0);\eta_+,\eta_-,\eta'_+,\eta'_-\rangle = \exp\left[\eta_+ D_+ + \eta_- D_- + \eta'_+ L_+ + \eta'_- L_-\right]$$
$$\times |(\omega,0),0,0\rangle, \tag{13.73}$$

where $\eta_+, \eta_-, \eta'_+, \eta'_-$ are real variables. All states of the representation $(\omega,0)$ can be obtained from (13.73), as shown schematically in Fig. 13.4.

Several parametrizations are again possible for the space $SO(4)/SO(2) \otimes SO(2)$. The algebraic parametrization is in terms of 4 real parameters $\beta_1, \beta_2, \alpha_1, \alpha_2$ appearing in the coherent state

$$|(\omega,0);\beta_1,\beta_2,\alpha_1,\alpha_2\rangle = \exp\left[\beta_1 X_1 + \beta_2 X_2 + \alpha_1 Y_1 + \alpha_2 Y_2\right]$$
$$\times |(\omega,0);0,0\rangle, \tag{13.74}$$

where X_1, X_2, Y_1, Y_2 are the real form of p. Instead of the four variables $\beta_1, \beta_2, \alpha_1, \alpha_2$, we can use the 'group' variables x_1, x_2, y_1, y_2, conveniently arranged in a 2×2 matrix

$$X = \begin{pmatrix} x_1 & y_1 \\ x_2 & y_2 \end{pmatrix}. \tag{13.75}$$

Introducing redundant variables such that

$$\begin{pmatrix} x_3 & y_3 \\ x_4 & y_4 \end{pmatrix} = \left[\begin{pmatrix} 1 & 0 \\ 0 & 1 \end{pmatrix} - \begin{pmatrix} x_1^2 + x_2^2 & x_1 y_1 + x_2 y_2 \\ x_1 y_1 + x_2 y_2 & y_1^2 + y_2^2 \end{pmatrix} \right]^{1/2} \equiv Y, \tag{13.76}$$

we can construct projective variables by

$$Z = \frac{X}{Y} = \begin{pmatrix} z_2^1 & z_1^2 \\ z_2^1 & z_2^2 \end{pmatrix}. \tag{13.77}$$

The Generic Coset $SO(n + 2)/SO(n) \otimes SO(2)$

A construction of the generic coset spaces $SO(n + 2)/SO(n) \otimes SO(2)$ can be done by generalizing the results of the previous subsection. The 'group' variables $x_1 x_2, \ldots, x_n; y_1, y_2, \ldots, y_n$ can be arranged into a $n \times 2$ matrix

$$\begin{pmatrix} x_1 & y_1 \\ x_2 & y_2 \\ \cdots & \cdots \\ x_n & y_n \end{pmatrix}. \tag{13.78}$$

The projective variables $z_i^1, z_i^2, i = 1, 2, .., n$, can be obtained from (13.78) by first introducing the $n \times 2$ matrix

$$Y = \left[I - \begin{pmatrix} \mathbf{x} \cdot \mathbf{x} & \mathbf{x} \cdot \mathbf{y} \\ \mathbf{x} \cdot \mathbf{y} & \mathbf{x} \cdot \mathbf{y} \end{pmatrix} \right]^{1/2}, \tag{13.79}$$

and then taking $Z = XY^{-1}$ to produce the $n \times 2$ matrix

$$\begin{pmatrix} z_1^1 & z_1^2 \\ z_2^1 & z_2^2 \\ \cdots & \cdots \\ z_n^1 & z_n^2 \end{pmatrix}. \tag{13.80}$$

13.8   Coset Spaces $SO(n, m)/SO(n) \otimes SO(m)$

These coset spaces can be treated in the same way as the spaces $SO(n+m)/SO(n) \otimes SO(m)$. Particularly important in applications are the cosets $SO(n, 1)/SO(n)$ which appear in relativistic quantum mechanics and relativistic quantum field theories.

13.8.1   The Coset Space $SO(2, 1)/SO(2)$: The Hyperboloid H^2

We consider here the coset space $SO(2, 1)/SO(2)$ with dimension 2. This space can also be mapped onto the coset spaces $U(1, 1)/U(1) \otimes U(1)$ and $SU(1, 1)/U(1)$.

The algebra $so(2, 1)$ is the complex extension of $so(3)$, as given in (1.18), and composed of $Y_1 = -iX_1$, $Y_2 = -iX_2$, $Y_3 = X_3$. The decomposition $g = h \oplus p$ is

$$g = so(2, 1) \doteq Y_1, Y_2, Y_3$$
$$h = so(2) \doteq Y_3$$
$$p \doteq Y_2, Y_3 = -iX_1, -iX_2. \tag{13.81}$$

Parametrizations of the Coset Space $SO(2, 1)/SO(2)$

The 'algebraic' parametrization of the coset space $SO(2, 1)/SO(2)$ is in terms of two parameters β_1, β_2. The 'group' parametrization can be obtained from β_1, β_2 with the change of variables

$$x_i = \beta_i \frac{\sinh \left(\sum_{i'} \beta_{i'}^2 \right)^{1/2}}{\left(\sum_{i'} \beta_{i'}^2 \right)^{1/2}}, \quad i, i' = 1, 2. \tag{13.82}$$

Introducing the redundant variable x_3

$$x_3 = \pm \sqrt{1 + x_1^2 + x_2^2}, \tag{13.83}$$

one has

$$x_3^2 - x_1^2 - x_2^2 = 1. \tag{13.84}$$

This space is the two-hyperboloid embedded in R_3, $H^2 \subset R_3$, shown in Fig. 13.5.

Points on the hyperboloid can be parametrized by two angles θ, ϕ through

$$x_1 = \sinh \theta \cos \phi \ , \ x_2 = \sinh \theta \sin \phi \ , \ x_3 = \cosh \theta. \tag{13.85}$$

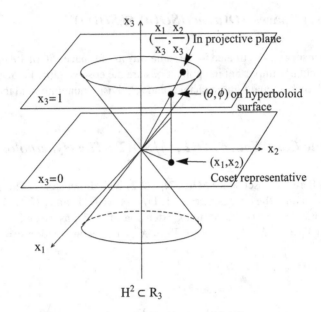

Fig. 13.5 Group variables x_1, x_2, x_3 with $x_3^2 - x_1^2 - x_2^2 = 1$ or (θ, ϕ) for the hyperboloid $H^2 \subset R_3$, and projective variables $\frac{x_1}{x_3}, \frac{x_2}{x_3}$

The projective parametrization is obtained by the change of variables

$$z_i = \frac{x_i}{\sqrt{1 + \sum_{i'} x_{i'}^2}}, \quad i, i' = 1, 2. \tag{13.86}$$

13.8.2   The Coset Space $SO(3,1)/SO(3)$

The Lorentz algebra $so(3,1)$ is the complex extension of $so(4)$, as given in (1.33). The decomposition $g = h \oplus p$ is

$$g = so(3,1) \doteq Y_1, Y_2, Y_3; L_1, L_2, L_3$$
$$h = so(3) \doteq L_1, L_2, L_3$$
$$p \doteq Y_1, Y_2, Y_3 = -iD_1, -iD_2, -iD_3. \tag{13.87}$$

Parametrizations of the Coset Space $SO(3,1)/SO(3)$

The 'algebraic' parametrization of this space is in terms of three variables $\beta_1, \beta_2, \beta_3$. The group parametrization is in terms of variables

$$x_i = \beta_i \frac{\sinh \left(\sum_{i'} \beta_{i'}^2\right)^{1/2}}{\left(\sum_{i'} \beta_{i'}^2\right)^{1/2}}, \quad i, i' = 1, 2, 3. \tag{13.88}$$

Adding a redundant variable $x_4 = \pm\sqrt{1 + x_1^2 + x_2^2 + x_3^2}$, we see that the group space is the three-hyperboloid, $H^3 \subset R_4$. This space can also be parametrized in terms of three angles θ, ϕ, ψ. The projective parametrization is in terms of the variables

$$z_i = \frac{x_i}{\sqrt{1 + \sum_{i'} x_{i'}^2}}, \quad i, i' = 1, 2, 3. \tag{13.89}$$

13.9   Action of the Coset $P = G/H$

In some applications, it is of interest to consider the action of the coset $P = G/H$. The coset P is

$$P = \exp p. \tag{13.90}$$

In the following subsections, the explicit form of the coset representative of P is given for some cases of practical interest. By applying the coset P to a point in the coset space, γ, we can reach other points, γ', in this space. The set of all points that can be reached by applying P is called the *orbit* of the point γ under G/H.

13.9.1   Cosets $SO(n+1)/SO(n)$

For these cosets, the algebraic parametrization is in terms of variables $\beta_i, i = 1, \ldots, n$, in R_n. The coset can be written as

$$\exp \begin{pmatrix} & & \beta_1 \\ & 0 & \cdots \\ & & \beta_n \\ -\beta_1 & \cdots & -\beta_n & 0 \end{pmatrix} = \exp \begin{pmatrix} 0 & \mathbf{B} \\ -\mathbf{B}^t & 0 \end{pmatrix}. \tag{13.91}$$

By expanding the exponential, one can rewrite (13.91) in terms of $(n+1) \times (n+1)$ matrices

$$\begin{pmatrix} \cos \sqrt{\mathbf{BB}^t} & \mathbf{B} \frac{\sin \sqrt{\mathbf{BB}^t}}{\sqrt{\mathbf{BB}^t}} \\ -\mathbf{B}^t \frac{\sin \sqrt{\mathbf{B}^t\mathbf{B}}}{\sqrt{\mathbf{B}^t\mathbf{B}}} & \cos \sqrt{\mathbf{B}^t\mathbf{B}} \end{pmatrix}. \tag{13.92}$$

Introducing the group variables x_i, $i = 1, \ldots, n$, one can rewrite the matrix (13.92) as

$$\begin{pmatrix} [\mathbf{I}_n - \mathbf{XX}^t]^{1/2} & \mathbf{X} \\ -\mathbf{X}^t & [\mathbf{I}_n - \mathbf{X}^t\mathbf{X}]^{1/2} \end{pmatrix}, \tag{13.93}$$

with

$$x_i = b_i \frac{\sin\left(\sum_{i'=1}^{n} \beta_{i'}^2\right)}{\left(\sum_{i'=1}^{n} \beta_{i'}^2\right)}, \tag{13.94}$$

or, introducing a redundant variable $x_{n+1} = \pm\sqrt{1 - \sum_i x_i^2}$, such that

$$x_1^2 + x_2^2 + \ldots + x_n^2 + x_{n+1}^2 = 1, \tag{13.95}$$

as

$$\begin{pmatrix} \cdot & \cdot & \cdot & x_1 \\ \cdot & \cdot & \cdot & \ldots \\ \cdot & \cdot & \cdot & x_n \\ \hline -x_1 & \ldots & -x_n & x_{n+1} \end{pmatrix}. \tag{13.96}$$

The coset space is here $S^n \subset R_{n+1}$.

Example 1. The coset $SO(3)/SO(2)$

In this case the coset representative is

$$\exp\begin{pmatrix} 0 & 0 & \beta_1 \\ 0 & 0 & \beta_2 \\ -\beta_1 & -\beta_2 & 0 \end{pmatrix} = \begin{pmatrix} \cdot & \cdot & x_1 \\ \cdot & \cdot & x_2 \\ \hline -x_1 & -x_2 & x_3 \end{pmatrix} \tag{13.97}$$

and the coset space is the sphere $S^2 \subset R_3$. By acting with $P = G/H$ on the coordinates of the north pole $(0, 0, 1)$ we can map it into the entire sphere

$$\begin{pmatrix} \cdot & \cdot & x_1 \\ \cdot & \cdot & x_2 \\ \hline -x_1 & -x_2 & x_3 \end{pmatrix}\begin{pmatrix} 0 \\ 0 \\ 1 \end{pmatrix} = \begin{pmatrix} x_1 \\ x_2 \\ x_3 \end{pmatrix}. \tag{13.98}$$

Therefore, the sphere S^2 is called the orbit of the north pole under $SO(3)/SO(2)$, Fig. 13.6.

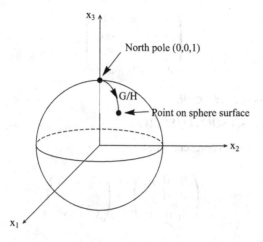

Fig. 13.6 The sphere S^2 as the orbit of the north pole $(0, 0, 1)$ under $SO(3)/SO(2)$

13.9.2   Cosets $SO(n, 1)/SO(n)$

For these cosets, one can repeat the treatment of the previous subsections with appropriate change of signs and replacement of trigonometric functions with hyperbolic functions. The cosets can be written as

$$\exp\left(\begin{array}{c|c} 0 & \mathbf{B} \\ \hline \mathbf{B}^t & 0 \end{array}\right). \tag{13.99}$$

By expanding the exponential, one obtains

$$\left(\begin{array}{c|c} \cosh\sqrt{\mathbf{BB}^t} & \mathbf{B}\dfrac{\sinh\sqrt{\mathbf{BB}^t}}{\sqrt{\mathbf{BB}^t}} \\ \hline \mathbf{B}^t\dfrac{\sinh\sqrt{\mathbf{B}^t\mathbf{B}}}{\sqrt{\mathbf{B}^t\mathbf{B}}} & \cosh\sqrt{\mathbf{B}^t\mathbf{B}} \end{array}\right). \tag{13.100}$$

Introducing the variables x_i, $i = 1, \ldots, n$, and the redundant variable $x_{n+1} = \pm\sqrt{1 + \sum_i x_i^2}$, such that

$$-x_1^2 - \ldots - x_{n-1}^2 - x_n^2 + x_{n+1}^2 = 1, \tag{13.101}$$

one can rewrite the coset as

$$\begin{pmatrix} \cdot & \cdot & \cdot & x_1 \\ \cdot & \cdot & \cdot & \cdots \\ \cdot & \cdot & \cdot & x_n \\ x_1 & \cdots & x_n & x_{n+1} \end{pmatrix}. \tag{13.102}$$

The coset space is here $H^n \subset R_{n+1}$.

Example 2. The coset $SO(2,1)/SO(2)$

In this case the coset representative is

$$\exp \begin{pmatrix} 0 & 0 & \beta_1 \\ 0 & 0 & \beta_2 \\ \beta_1 & \beta_2 & 0 \end{pmatrix} = \begin{pmatrix} \cdot & \cdot & x_1 \\ & & x_2 \\ x_1 & x_2 & x_3 \end{pmatrix}. \tag{13.103}$$

This space is $H^2 \subset R_3$.

Additional properties of the coset spaces, such as geodesics, distance, metric and volume of cosets are discussed in Gilmore (1974).

Chapter 14
Spectrum Generating Algebras and Dynamic Symmetries

14.1 Spectrum Generating Algebras

One of the most important applications of Lie algebras has been to the study of physical systems for which the Hamiltonian H and other operators of physical interest T can be written in terms of elements G_α of a Lie algebra g,

$$H = f(G_\alpha) \quad G_\alpha \in g \tag{14.1}$$

and

$$T = t(G_\alpha) \quad G_\alpha \in g. \tag{14.2}$$

The Lie algebra g is then called the Spectrum Generating Algebra (SGA) of the problem. The functionals f and t are usually polynomials in the elements G_α, although cases have been studied in which $1/H$ rather than H is a polynomial in the elements of the algebra (the so-called Coulomb problem discussed in Chap. 15). It turns out that most many-body problems, that is Hamiltonian problems for many interacting particles, can be cast in the form of a polynomial expansion in G_α. The method is thus particularly important for this case. It is convenient to use the double index notation, $G_{\alpha\beta}$, of Chaps. 9 and 10. The expansion can then be written as

$$H = E_0 + \sum_{\alpha\beta} \varepsilon_{\alpha\beta} G_{\alpha\beta} + \frac{1}{2} \sum_{\alpha\beta\gamma\delta} u_{\alpha\beta\gamma\delta} G_{\alpha\beta} G_{\gamma\delta} + \ldots \tag{14.3}$$

The term linear in the elements is called one-body term, the term quadratic in the elements is called two-body term, etc.. Most theories stop the expansion at two-body terms, although in some cases three- and higher-body terms have been considered.

© Springer-Verlag Berlin Heidelberg 2015

F. Iachello, *Lie Algebras and Applications*, Lecture Notes in Physics 891,
DOI 10.1007/978-3-662-44494-8_14

The term E_0 is an overall constant, which sets the zero of the energy, and the coefficients $\varepsilon_{\alpha\beta}, u_{\alpha\beta\gamma\delta}, \ldots$ depend on the physical system under consideration.

14.2 Dynamic Symmetries

A particularly interesting situation occurs when the Hamiltonian H does not contain all elements of g, but only those combinations which form the Casimir operators of a chain of algebras originating from $g \supset g' \supset g'' \supset \ldots$

$$H = f(C_i). \tag{14.4}$$

For these situations, called dynamic symmetries (DS), the eigenvalue problem for H can be solved in explicit analytic form, since the Casimir operators are diagonal in the basis provided by $g \supset g' \supset g'' \supset \ldots$. The energy eigenvalues are given by a formula called *energy formula* in terms of the quantum numbers that characterize the representations of $g \supset g' \supset \ldots$, while the matrix elements of the transition operators T are given in terms of isoscalar factor of the chain $g \supset g' \supset g'' \supset \ldots$. When the Hamiltonian H is linear in the Casimir operators

$$H = E_0 + \alpha C(g) + \alpha' C(g') + \alpha'' C(g'') + \ldots \tag{14.5}$$

the energy formula is simply

$$E = \langle H \rangle = E_0 + \alpha \langle C(g) \rangle + \alpha' \langle C(g') \rangle + \alpha'' \langle C(g'') \rangle + \ldots, \tag{14.6}$$

where $\langle \rangle$ denotes expectation value in the appropriate representation. Several examples will be discussed in the following sections. In these examples, taken from the fields of molecular physics, nuclear physics and hadronic physics, the Hamiltonian H is at most quadratic in the elements of the algebra g, and hence the Hamiltonian with dynamic symmetry contains at most Casimir operators of order 2. Systems for which the Hamiltonian has a dynamic symmetry are also called *exactly solvable problems*. Dynamic symmetries were introduced in Barut and Böhm (1965) and Dothan et al. (1965) and subsequently exploited in nuclear physics (Iachello 1979, p. 420) and other areas.

14.3 Bosonic Systems

Consider a system composed of N bosons $b_\alpha (\alpha = 1, 2, \ldots, n)$. The (number conserving) Hamiltonian operator for this system can be written as

$$H = E_0 + \sum_{\alpha\beta} \tilde{\varepsilon}_{\alpha\beta} b_\alpha^\dagger b_\beta + \frac{1}{2} \sum_{\alpha\beta\gamma\delta} u_{\alpha\beta\gamma\delta} b_\alpha^\dagger b_\beta^\dagger b_\gamma b_\delta + \dots . \tag{14.7}$$

After rearrangement of some of the boson operators, this Hamiltonian can be written as above with $G_{\alpha\beta} = b_\alpha^\dagger b_\beta$ and

$$\varepsilon_{\alpha\beta} = \tilde{\varepsilon}_{\alpha\beta} - \frac{1}{2} \sum_\gamma u_{\alpha\gamma\gamma\beta} . \tag{14.8}$$

The bilinear products $b_\alpha^\dagger b_\beta$ are the elements of the Lie algebra $u(n)$. Hence $u(n)$ is the spectrum generating algebra of this problem. The transition operators

$$T = \sum_{\alpha\beta} t_{\alpha\beta} b_\alpha^\dagger b_\beta + \dots \tag{14.9}$$

can also be written in terms of the elements of the Lie algebra $u(n)$. A physical system is characterized by a set of parameters $\tilde{\varepsilon}_{\alpha\beta}, u_{\alpha\beta\gamma\delta}$ and $t_{\alpha\beta}$. The methods of the previous chapters can then be used to solve the eigenvalue problem for H, that is to find the energy spectrum of the system under consideration, and to calculate matrix elements of operators.

Dynamic symmetries of this system can be studied by breaking $u(n)$ in all possible ways. Often additional conditions are imposed on the breaking of the algebra g into its subalgebras g'. In most applications, the system under study is rotationally invariant. Hence the algebra $so(3)$ must be included in the chain $g \supset g' \supset g'' \supset \dots$. The breakings of $u(n)$ into its subalgebras containing $so(3)$ were enumerated in Chap. 9. Knowing these breakings one can construct the corresponding dynamic symmetries. Two examples of boson dynamic symmetries will be given here.

14.3.1   Dynamic Symmetries of $u(4)$

Consider the algebra $g \equiv u(4)$ discussed in Chap. 9. This algebra has two subalgebra chains containing $so(3)$

$$u(3) \supset so(3) \supset so(2) \quad (I)$$

$$u(4) \Big\langle \tag{14.10}$$

$$so(4) \supset so(3) \supset so(2) \quad (II)$$

and correspondingly two possible dynamic symmetries.

Dynamic symmetry I

The Hamiltonian in this case can be written as

$$H^{(I)} = E_0 + \varepsilon C_1(u(3)) + \alpha C_2(u(3)) + \beta C_2(so(3)). \tag{14.11}$$

The Casimir operator of $so(2)$ could be added to this Hamiltonian. However, this corresponds physically to placing the system in an external field that splits the degeneracy of the angular momentum. In the absence of external fields, the Casimir operator of $so(2)$ can be deleted. Also the invariant operators of $u(4)$ could be included in the overall constant E_0,

$$E_0 = E_{00} + E_{01}C_1(u(4)) + E_{02}C_2(u(4)). \tag{14.12}$$

In the representation $[N]$ of $u(4)$ appropriate to bosons,

$$E_0 = E_{00} + E_{01}N + E_{02}N(N+3) \tag{14.13}$$

is a constant for all states in the basis.

The eigenvalues of $H^{(I)}$ in the basis $\mid N, n_p, L, M_L\rangle$ can simply be found from the eigenvalues of the Casimir operators given in Chap. 7,

$$E^{(I)}(N, n_p, L, M_L) = E_0 + \varepsilon n_p + \alpha n_p(n_p + 2) + \beta L(L+1). \tag{14.14}$$

Usually, a hierarchy of couplings occurs, in such a way that successive splittings are smaller and smaller, $|\varepsilon| \gg |\alpha| \gg |\beta|$. It is customary to show the spectrum in an *energy level diagram*. The energy level diagram of the dynamic symmetry I of $u(4)$, with $\varepsilon > 0$, is shown in the top part of Fig. 14.1.

This is the spectrum of a truncated three dimensional anharmonic oscillator. When $\alpha = \beta = 0$, the spectrum is called harmonic. The spectrum is truncated, since according to the branching rules of Chap. 9, $n_p \leq N$.

Dynamic symmetry II

The Hamiltonian for this case is

$$H^{(II)} = E_0 + AC_2(so(4)) + BC_2(so(3)), \tag{14.15}$$

where again the Casimir operator of $so(2)$ has been deleted. The eigenvalues of $H^{(II)}$ in the basis $|N, \omega, L, M_L\rangle$ are

$$E^{(II)}(N, \omega, L, M_L) = E_0 + A\omega(\omega + 2) + BL(L+1). \tag{14.16}$$

Usually $|A| \gg |B|$. The energy level diagram of this dynamic symmetry, when $A < 0$, is shown in the bottom portion of Fig. 14.1. It represents the spectrum of the truncated three dimensional rotovibrator. In the figure, the usual vibrational quantum number is also shown. A three-dimensional rotovibrator is an object, such as a diatomic molecule, which can vibrate and rotate around an axis perpendicular to

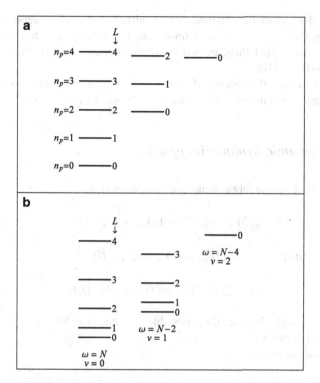

Fig. 14.1 The energy level diagram of the dynamic symmetries of $u(4)$: **a** $u(4) \supset u(3) \supset so(3) \supset so(2)$; **b** $u(4) \supset so(4) \supset so(3) \supset so(2)$. The representation [4] is shown

Fig. 14.2 A diatomic molecule is shown as an example of a truncated three-dimensional rotovibrator

the line joining the two atoms, as shown in Fig. 14.2. This spectrum is characterized by a set of states belonging to the same representation of $so(4)$, called a rotational band. Different representations of $so(4)$ correspond to different vibrational excitations. The spectrum is truncated since, according to the branching rules of Chap. 9, $\omega \leq N$.

The two dynamic symmetries represent special situations. In general, the Hamiltonian H will contain Casimir operators of both chains

$$H = E_0 + \varepsilon C_1(u(3)) + \alpha C_2(u(3)) + A C_2(so(4)) + B C_2(so(3)). \tag{14.17}$$

The eigenvalues must be obtained numerically. For this most general case, Lie algebraic methods provide a basis upon which the numerical diagonalization is done. In the case of $u(4)$, there are two such bases, corresponding to the two dynamic symmetries (I) and (II).

Models based on the algebra of $u(4)$ and its dynamic symmetries have found many applications in molecular physics (Iachello and Levine 1995).

14.3.2 Dynamic Symmetries of $u(6)$

For $g \equiv u(6)$, the subalgebra chains containing $so(3)$ are

$$
\begin{array}{c}
u(5) \supset so(5) \supset so(3) \supset so(2) \quad (I) \\[4pt]
u(6) - \quad su(3) \supset so(3) \supset so(2) \quad (II) \\[4pt]
so(6) \supset so(5) \supset so(3) \supset so(2) \quad (III)
\end{array}
\qquad (14.18)
$$

and correspondingly there are three possible dynamic symmetries.

Dynamic symmetry I
The Hamiltonian is

$$
H^{(I)} = E_0 + \varepsilon C_1(u(5)) + \alpha C_2(u(5)) + \beta C_2(so(5)) + \gamma C_2(so(3)), \qquad (14.19)
$$

where again the Casimir operator of $so(2)$ has been omitted. The eigenvalues in the basis $|N, n_d, v, n_\Delta, L, M_L\rangle$ are

$$
\begin{aligned}
E^{(I)}(N, n_d, v, n_\Delta, L, M_L) = {}& E_0 + \varepsilon n_d + \alpha n_d (n_d + 4) \\
& + \beta v(v + 3) + \gamma L(L + 1). \qquad (14.20)
\end{aligned}
$$

Usually $|\varepsilon| \gg |\alpha| \gg |\beta| \gg |\gamma|$. The energy level diagram, when $\varepsilon > 0$, is shown in Fig. 14.3 (top panel). It represents the spectrum of the truncated five-dimensional anharmonic oscillator. If $\alpha = \beta = \gamma = 0$, the spectrum is called harmonic.

Dynamic symmetry II
The Hamiltonian is

$$
H^{(II)} = E_0 + \kappa C_2(su(3)) + \kappa' C_2(so(3)). \qquad (14.21)
$$

Its eigenvalues in the basis $|N, \lambda, \mu, K, L, M_L\rangle$ are

$$
E^{(II)}(N, \lambda, \mu, K, L, M_L) = E_0 + \kappa(\lambda^2 + \mu^2 + \lambda\mu + 3\lambda + 3\mu) + \kappa' L(L + 1). \qquad (14.22)
$$

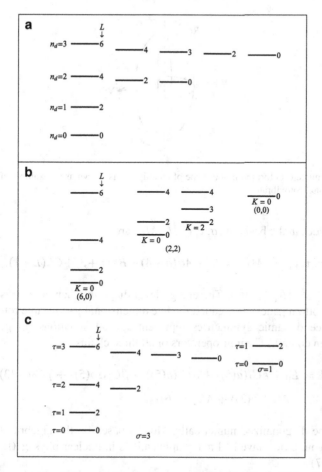

Fig. 14.3 The energy level diagram of the dynamic symmetries of $u(6)$: **a** $u(6) \supset u(5) \supset so(5) \supset so(3) \supset so(2)$; **b** $u(6) \supset su(3) \supset so(3) \supset so(2)$; **c** $u(6) \supset so(6) \supset so(5) \supset so(3) \supset so(2)$. The representation [3] is shown

Usually $|\kappa| \gg |\kappa'|$. The corresponding energy level diagram, when $\kappa < 0$, is shown in Fig. 14.3 (middle panel). This is the spectrum of the truncated five-dimensional rotovibrator. A five-dimensional rotovibrator is an object, such as an atomic nucleus deformed in the shape of a quadrupole, which can vibrate and rotate around an axis perpendicular to the symmetry axis of the quadrupole, as shown in Fig. 14.4.

Each representation (λ, μ) represents a rotational band. Different representations (λ, μ) label vibrational excitations.

Dynamic symmetry III
The Hamiltonian is

$$H^{(III)} = E_0 + AC_2(so(6)) + BC_2(so(5)) + CC_2(so(3)). \qquad (14.23)$$

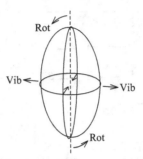

Fig. 14.4 A nucleus deformed in the shape of an ellipsoid is shown as an example of a truncated five-dimensional rotovibrator

Its eigenvalues in the basis $|N, \sigma, \tau, \nu_\Delta, L, M_L\rangle$ are

$$E^{(III)}(N, \sigma, \tau, \nu_\Delta, L, M_L) = E_0 + A\sigma(\sigma+4) + B\tau(\tau+3) + CL(L+1). \quad (14.24)$$

Usually $|A| \gg |B| \gg |C|$. The energy level diagram when $A < 0$ is shown in Fig. 14.3 (bottom panel). It is called the five dimensional γ−unstable rotovibrator.

The three dynamic symmetries represent special situations. In general, the Hamiltonian contains Casimir operators of all three chains

$$H = E_0 + \varepsilon C_1(u(5)) + \alpha C_2(u(5)) + \beta C_2(so(5)) + \gamma C_2(so(3))$$
$$+\kappa C_2(su(3)) + A C_2(so(6)) \quad (14.25)$$

and must be diagonalized numerically. Models based on the algebra $u(6)$ and its dynamic symmetries have had many applications in nuclear physics (Iachello and Arima 1987).

14.4 Fermionic Systems

Spectrum generating algebras and dynamic symmetries can also be used for fermions. Consider a system composed of N_F fermions $a_i (i = 1, 2, .., n)$. The number conserving Hamiltonian operator for this system can be written as

$$H = E_0 + \sum_{ii'} \tilde{\eta}_{ii'} a_i^\dagger a_{i'} + \frac{1}{2} \sum_{ii'kk'} v_{ii'kk'} a_i^\dagger a_{i'}^\dagger a_k a_{k'} + ... \quad (14.26)$$

After rearrangement of some of the fermion operators, the Hamiltonian can be rewritten as

$$H = E_0 + \sum_{ii'} \eta_{ii'} G_{ii'} + \frac{1}{2} \sum_{iki'k'} v_{ii'kk'} G_{ik} G_{i'k'} + \dots . \tag{14.27}$$

The elements $G_{ii'}$ span the Lie algebra $u(n)$. Hence $u(n)$ is the spectrum generating algebra of this problem. All operators can be expanded into elements of this algebra. The transition operators

$$T = \sum_{ii'} t_{ii'} a_i^\dagger a_{i'} + \dots \tag{14.28}$$

are written as

$$T = \sum_{ii'} t_{ii'} G_{ii'} + \dots . \tag{14.29}$$

A physical system is characterized by a set of parameters $\eta_{ii'}$, $v_{ii'kk'}$ and $t_{ii'}$. One can use the methods of the previous chapters to solve the eigenvalue problem for H and to calculate the matrix elements of operators.

Dynamic symmetries of these systems can be studied by breaking $u(n)$ in all possible ways. There are two types of problems, the atomic and nuclear many body problem in which the Hamiltonian is rotationally invariant and thus $so(3)$ must be included in the chain $g \supset g' \supset g' \supset \dots$, and other problems in which one treats internal degrees of freedom and thus rotationally invariance need not to be imposed. The former problem is identically to that discussed in Sect. 14.3. In the following subsections two examples will be given of the latter problem.

14.4.1 Dynamic Symmetry of $u(4)$

The Wigner algebra $u(4)$ of Sect. 10.6.1 is described by the chain

$$u(4) \supset su(4) \supset su_T(2) \oplus su_S(2) \supset spin_T(2) \oplus spin_S(2). \tag{14.30}$$

The basis states can be denoted by $|(P, P', P''); T, S; M_T, M_S\rangle$. The Hamiltonian with dynamic symmetry is

$$H = E_0 + a C_2(su(4)) + b C_2(su_T(2)) + c C_2(su_S(2)) + d C_1(spin_T(2)). \tag{14.31}$$

In this Hamiltonian, the Casimir operator of $spin_S(2)$ is not included, unless the system is placed in an external field that splits the $su_S(2)$ degeneracy, but the Casimir operator of $spin_T(2)$ is included. This group acts on an abstract isotopic spin space. In this space, there are interactions (for example the electromagnetic interaction) that split the $su_T(2)$ degeneracy. The eigenvalues of the Hamiltonian are

$$E(P, P', P''; T, S; M_T, M_S) = E_0 + a(P^2 + 4P + P'^2 + P''^2)$$

$$+bT(T+1) + cS(S+1) + dM_T. \quad (14.32)$$

A simple case is when $b = c = d = 0$ (Franzini and Radicati 1963).

14.4.2 Dynamic Symmetry of $u(6)$

The Gürsey-Radicati $u(6)$ of Sect. 10.6.2 is described by the chain

$$u(6) \supset su(6) \supset su_F(3) \oplus su_S(2) \supset su_T(2) \oplus u_Y(1) \oplus su_S(2)$$

$$\supset spin_T(2) \oplus u_Y(1) \oplus spin_S(2). \quad (14.33)$$

The basis states are denoted by $\left| \dim[\lambda];^{\dim S} \dim[\mu_1, \mu_2]; T, Y, S; M_T, M_S \right\rangle$. For applications to elementary particle physics, the mass squared operator M^2 rather than the Hamiltonian H is expanded into the elements of a Lie algebra. The mass squared operator, M^2, with dynamic symmetry is written in terms of Casimir invariants as

$$M^2 = M_0^2 + aC_2(su_F(3)) + a'C_1(u_Y(1)) + bC_2(su_T(2)) + b'(C_1(u_Y(1)))^2$$

$$+cC_2(su_S(2)) + dC_1(spin_T(2)) + d'\,(C_1(spin_T(2)))^2\,, \quad (14.34)$$

with eigenvalues

$$M^2([\lambda], [\mu_1, \mu_2]; T, Y, S; M_T, M_S) = M_0^2 + a\,\langle C_2(su_F(3))\rangle + a'Y + bT(T+1)$$

$$+b'Y^2 + cS(S+1) + dM_T + d'M_T^2. \quad (14.35)$$

In the expansion (14.34), terms containing the Casimir operators of $spin_S(2)$ have been omitted, since the inclusion of these terms will amount to placing the system in an external field. Also, the invariant operators of $u(6)$ could be included in the overall constant M_0^2,

$$M_0^2 = M_{00}^2 + M_{01}C_1(u(6)) + M_{02}C_2(u(6)). \quad (14.36)$$

In the representation $[\lambda]$ of $u(6)$, the expectation value of M_0^2 is

$$M_0^2 = M_{00}^2 + M_{01}\,\langle C_1(u(6))\rangle + M_{02}\,\langle C_2(u(6))\rangle\,, \quad (14.37)$$

which is an overall constant. Furthermore, the terms containing the Casimir operators of $spin_T(2)$ describe electromagnetic splittings between states with different values of M_T. It turns out that these splittings are small, and therefore the terms d, d' can be omitted. Also, a dynamical input suggests that $b' = -\frac{1}{4}b$, thus yielding

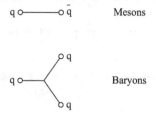

q o———o q̄   Mesons

q o——< o q
          o q   Baryons

Fig. 14.5 Mesons and baryons in the quark model

$$M^2\left([\lambda], [\mu_1, \mu_2]; T, Y, S; M_T, M_S\right) = M_0^2 + a\langle C_2(su_F(3))\rangle + a'Y$$
$$+b\left[T(T+1) - \frac{1}{4}Y^2\right] + cS(S+1).$$
$$(14.38)$$

This formula, called a *mass formula*, gives the masses of all states (particles) belonging to a given representation $[\lambda]$ of $su(6)$. It has been used in this form, M^2, or in its linear form, M, to describe the masses of particles composed of u, d, s quarks in the quark model. In this model, denoting by q a quark and by $\bar{q}$ an antiquark, particles called mesons are bound states of a quark an antiquark, $q\bar{q}$, while particles called baryons are bound states of there quarks, q^3, as schematically shown in Fig. 14.5.

Since quarks are assigned to the representation $[1, 0, 0, 0, 0]$ of $su(6)$, with dimension 6, Table 10.6, and antiquarks to the representation $[1, 1, 1, 1, 1]$, with dimension $\bar{6}$, the meson states belong to the representations

$$[1, 0, 0, 0, 0] \otimes [1, 1, 1, 1, 1] = [2, 1, 1, 1, 1] \oplus [0],$$
$$6 \otimes \bar{6} = 35 \oplus 1. \qquad (14.39)$$

Similarly, baryons belong to the representations

$$[1, 0, 0, 0, 0] \otimes [1, 0, 0, 0, 0] \otimes [1, 0, 0, 0, 0] = [3, 0, 0, 0, 0] \oplus [2, 1, 0, 0, 0]$$
$$\oplus [2, 1, 0, 0, 0] \oplus [1, 1, 1, 0, 0, 0]$$

$$6 \otimes 6 \otimes 6 = 56 \oplus 70 \oplus 70 \oplus 20. \qquad (14.40)$$

The mass level diagram for the representation $[3]$ of baryons is shown in Fig. 14.6, where it is divided into the two representations $^2 8$ and $^4 10$ of $su_S(2) \oplus su_F(3)$ (Gürsey and Radicati 1964).

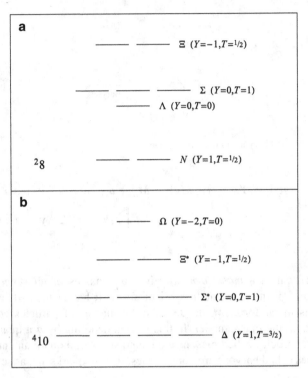

Fig. 14.6 The mass level diagram of the representation [3] of $u(6)$ with dimension 56: **a** Baryon octet, $^{2}8$; **b** baryon decuplet $^{4}10$. States are labeled by quantum numbers and by names of particles to which they correspond

The $su_F(3)$ flavor part of this formula, in its linear form

$$M = M_0 + a'Y + b\left[T(T+1) - \frac{1}{4}Y^2\right],\qquad(14.41)$$

called the Gell-Mann and Okubo *mass formula*, was the first explicit example of dynamic symmetry in physics (Gell-Mann 1962; Okubo 1962). Further details on applications of group theoretcial methods in particle physics can be found in (Georgi, 1962) and (Lipkin, 1966).

Chapter 15
Degeneracy Algebras and Dynamical Algebras

15.1 Degeneracy Algebras

Another important application of algebraic methods in physics is to the study of exactly solvable problems in quantum mechanics. Consider quantum mechanics in v dimensions described by the Hamiltonian

$$H = -\frac{\hbar^2}{2m}\nabla^2 + V(\mathbf{r}) \tag{15.1}$$

where ∇^2 is the Laplace operator and $\mathbf{r} \equiv (x_1, x_2, \ldots, x_v)$ denotes a vector in v dimensions with components $x_1, x_2, \ldots, x_v$. (In this chapter, v denotes the dimension of the space and n the so-called principal quantum number.) This Hamiltonian is obtained from the classical Hamiltonian

$$H = \frac{\mathbf{p}^2}{2m} + V(\mathbf{r}) \tag{15.2}$$

by the usual quantization procedure $\mathbf{p} \to \frac{\hbar}{i}\nabla$. If the Hamiltonian (15.1) can be written in terms of the Casimir operator C of an algebra g,

$$H = f(C) \tag{15.3}$$

the eigenvalue problem for H can be solved in explicit analytic form,

$$E = \langle f(C) \rangle. \tag{15.4}$$

This situation is a dynamic symmetry, Chap. 14, Sect. 14.2, except that only the Casimir operator of g and not those of the subalgebra chain $g \supset g' \supset g'' \supset \ldots$ appears in (15.3). The representations $[\lambda]$ of g still label the eigenstates of the

© Springer-Verlag Berlin Heidelberg 2015
F. Iachello, *Lie Algebras and Applications*, Lecture Notes in Physics 891,
DOI 10.1007/978-3-662-44494-8_15

Hamiltonian and the symbol $\langle\rangle$ denotes expectation value in the representation $[\lambda]$. If $\dim[\lambda] \neq 1$, more than one state has energy E. The state is said to be degenerate and the algebra g is called the *degeneracy algebra*, g_c, of the problem.

15.2   Degeneracy Algebras in $\nu \geq 2$ Dimensions

A particularly interesting class of problems is that of quantum mechanics in $\nu \geq 2$ dimensions with rotationally invariant potentials, $V = V(r)$. Here

$$r = (x_1^2 + x_2^2 + \ldots + x_\nu^2)^{1/2}. \tag{15.5}$$

This problem admits two and only two *exactly solvable* cases, the isotropic harmonic oscillator with $V(r) = \frac{1}{2}kr^2$, and the Coulomb (or Kepler) problem with $V(r) = \frac{k}{r}$.

15.2.1   The Isotropic Harmonic Oscillator

The Hamiltonian operator, in units where $\hbar = m = 1$ and $k = 1$, is

$$H = \frac{1}{2}(p^2 + r^2) = \frac{1}{2}(-\nabla^2 + r^2). \tag{15.6}$$

Introducing the bosonic realization of Chap. 9 written in differential form and generalizing the results of Example 3 one can write the linear Casimir operator of $u(\nu)$ as

$$C_1(u(\nu)) = \frac{1}{2}\sum_{j=1}^{\nu}\left(x_j - \frac{\partial}{\partial x_j}\right)\left(x_j + \frac{\partial}{\partial x_j}\right). \tag{15.7}$$

The basic commutation relations

$$\left[x_i, \frac{\partial}{\partial x_j}\right] = -\delta_{ij} \tag{15.8}$$

give

$$H = C_1(u(\nu)) + \frac{\nu}{2}. \tag{15.9}$$

The degeneracy algebra of the ν-dimensional harmonic oscillator is thus $u(\nu)$. Jauch and Hill (1940) States are characterized by the totally symmetric irreducible representations $[n, 0, \ldots, 0] \equiv [n]$ of $u(\nu)$, with eigenvalues

$$E(n) = n + \frac{\nu}{2} \qquad n = 0, 1, \ldots, \infty. \tag{15.10}$$

Although harmonic oscillator problems are best attacked by bosonic realizations of Lie algebras, it is still of interest to consider differential realizations in terms of coordinates $\mathbf{r} \equiv (x_1, x_2, \ldots, x_\nu)$ and momenta $\mathbf{p} = \frac{1}{i}\nabla \equiv \frac{1}{i}\left(\frac{\partial}{\partial x_1}, \frac{\partial}{\partial x_2}, \ldots, \frac{\partial}{\partial x_\nu}\right)$.

Example 1. Isotropic harmonic oscillator in three dimensions

As discussed in Chap. 9, this problem is best solved in spherical coordinates r, ϑ, φ. The Hamiltonian (15.2) is

$$H = \frac{\mathbf{p}^2}{2} + \frac{\mathbf{r}^2}{2}, \tag{15.11}$$

where $\mathbf{r}$ and $\mathbf{p}$ are here three-dimensional vectors. The nine operators

$$H = \frac{1}{2}\left(\mathbf{r}^2 + \mathbf{p}^2\right)$$

$$\mathbf{L} = -i\sqrt{2}[\mathbf{r} \times \mathbf{p}]^{(1)}$$

$$\mathbf{Q} = [\mathbf{r} \times \mathbf{r} + \mathbf{p} \times \mathbf{p}]^{(2)}, \tag{15.12}$$

where $\mathbf{L}$ and $\mathbf{Q}$ are the angular momentum vector and quadrupole (rank-2) tensor in three dimensions, satisfy commutation relations isomorphic to those of $u(3)$. (The notation used here is that of De Shalit and Talmi (1963), where $\mathbf{r}$ and $\mathbf{p}$ are rank-1 tensors, and the superscript denotes tensor couplings. The operator $\mathbf{Q}$ has an additional factor, often $\sqrt{8}$, in other definitions.) The algebra $u(3)$ has three invariant Casimir operators, C_1, C_2 and C_3. The Hamiltonian H of (15.11) can be rewritten as

$$H = C_1(u(3)) + \frac{3}{2}. \tag{15.13}$$

(Note that H commutes with the eight elements of the algebra $su(3)$, $\mathbf{L}$ and $\mathbf{Q}$.) The eigenstates are labelled by the irreducible representations $[n, 0, 0] \equiv [n]$ of $u(3)$. A complete labelling is $|n, l, m\rangle$ with

$$\left| \begin{matrix} u(3) \supset so(3) \supset so(2) \\ \downarrow \qquad \downarrow \qquad \downarrow \\ n \qquad l \qquad m \end{matrix} \right). \tag{15.14}$$

The branching of the representations $[n]$ of $u(3)$ into representations of $so(3) \supset so(2)$ has been discussed in Chap. 9. The values of l for each n are given by $l = n, n-2, \ldots, 1$ or 0 ($n =$ odd or even). The values of m are $m = -l, \ldots, +l$. The quantum number $n = 0, \ldots, \infty$. The energy level diagram

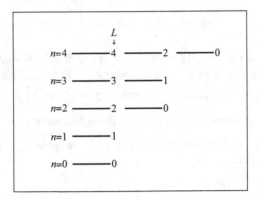

Fig. 15.1 The spectrum of the three-dimensional harmonic oscillator. States up to $n = 4$ are shown

$$E(n) = n + \frac{3}{2} \tag{15.15}$$

for this case is shown in Fig. 15.1. The three dimensional harmonic oscillator finds useful applications in a variety of problems in physics.

Example 2. Isotropic harmonic oscillator in five dimensions

This problem is of interest in nuclear physics, in the study of quadrupole oscillations of a liquid drop. The problem is best solved by introducing coordinates $\alpha_\mu(\mu = 0, \pm 1, \pm 2)$ and momenta $\pi_\mu(\mu = 0, \pm 1, \pm 2)$, called Bohr variables. The coordinates α_μ are the five components of a quadrupole (rank-2) tensor with respect to rotations. The Hamiltonian in dimensionless units is

$$H = \frac{1}{2} \left(\sum_\mu \pi_\mu^2 + \sum_\mu \alpha_\mu^2 \right). \tag{15.16}$$

This Hamiltonian can be rewritten as

$$H = C_1(u(5)) + \frac{5}{2}. \tag{15.17}$$

The basis states are labelled by the totally symmetric representations $[n, 0, 0, 0, 0] \equiv [n]$ of $u(5)$. The complete labelling is $|n, \tau, \nu_\Delta, l, m\rangle$ with

$$\left| \begin{array}{cccc} u(5) \supset so(5) \supset so(3) \supset so(2) \\ \downarrow \quad\quad \downarrow \quad\quad \downarrow \quad\quad \downarrow \\ n \quad\quad \tau, \nu_\Delta \quad\quad l \quad\quad m \end{array} \right\rangle. \tag{15.18}$$

The branching of the representations $[n]$ of $u(5)$ is given in Chap. 9, Sect. 9.6. The energy level diagram for this case

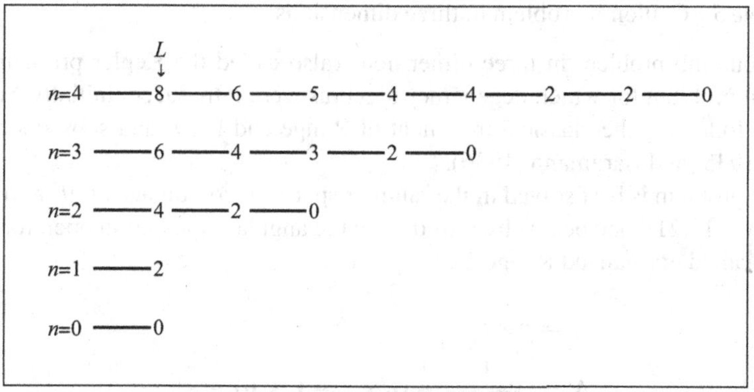

Fig. 15.2 The spectrum of the five-dimensional harmonic oscillator. States up to $n = 4$ are shown

$$E(n) = n + \frac{5}{2} \tag{15.19}$$

is given in Fig. 15.2.

15.2.2   The Coulomb Problem

The Hamiltonian operator for this problem, when $\hbar = m = 1$ and $k = 1$, is

$$H = \frac{p^2}{2} - \frac{1}{r} = -\frac{\nabla^2}{2} - \frac{1}{r}. \tag{15.20}$$

While the derivation of (15.13) is straightforward, the rewriting of H in terms of Casimir operators here is more involved (see Examples 3 and 4). For bound states, where $E = \langle H \rangle < 0$, the Hamiltonian H can be rewritten as

$$H = -\frac{1}{2\left[C_2(so(\nu + 1)) + \left(\frac{\nu-1}{2}\right)^2\right]}. \tag{15.21}$$

The degeneracy algebra of the ν-dimensional Coulomb problem is thus $so(\nu + 1)$. (The Coulomb problem is a case in which the Hamiltonian H is not linear in the Casimir operators, but rather its inverse, $1/H$, is.) The eigenstates are characterized by the totally symmetric irreducible representations $[\omega, 0, 0, \ldots, 0] \equiv [\omega]$ of $so(\nu + 1)$ with eigenvalues

$$E(\omega) = -\frac{1}{2\left[\omega(\omega + \nu - 1) + \left(\frac{\nu-1}{2}\right)^2\right]} \qquad \omega = 0, 1, \ldots, \infty. \tag{15.22}$$

Example 3. Coulomb problem in three dimensions

The Coulomb problem in three dimensions (also called the Kepler problem) was the first problem for which degeneracy algebras were introduced, initially by Pauli (1926) following the classical treatment of Runge and Lenz, and subsequently by Fock (1935) and Bargmann (1936).

The problem is best solved in the familiar spherical coordinates, r, ϑ, φ. In order to derive (15.21), one begins by introducing the angular momentum operator $\mathbf{L}$ and the so-called normalized Runge-Lenz vector $\mathbf{A}$

$$\mathbf{L} = \mathbf{r} \times \mathbf{p}$$

$$\mathbf{A} = \frac{1}{\sqrt{-2H}} [\frac{1}{2} (\mathbf{p} \times \mathbf{L} - \mathbf{L} \times \mathbf{p}) - \frac{\mathbf{r}}{r}]. \tag{15.23}$$

Using the commutation relations of $\mathbf{r}$ and $\mathbf{p} = \frac{1}{i}\nabla$, one can show that the six components L_i ($i = 1, 2, 3$) and A_i ($i = 1, 2, 3$) satisfy, when $E < 0$, the commutation relations of the Lie algebra $so(4)$

$$\left[L_i, L_j\right] = i\varepsilon_{ijk}L_k$$

$$\left[L_i, A_j\right] = i\varepsilon_{ijk}A_k$$

$$\left[A_i, A_j\right] = i\varepsilon_{ijk}L_k \quad, \tag{15.24}$$

while when $E > 0$ they satisfy the commutation relations of the non-compact algebra $so(3, 1)$

$$\left[L_i, L_j\right] = i\varepsilon_{ijk}L_k$$

$$\left[L_i, A_j\right] = i\varepsilon_{ijk}A_k$$

$$\left[A_i, A_j\right] = -i\varepsilon_{ijk}L_k \quad. \tag{15.25}$$

The components of the angular momentum and Runge-Lenz vector are thus, when $E < 0$, elements of the Lie algebra $so(4)$. This algebra has two quadratic Casimir operators, C_2 and C_2' that can be written as

$$C_2(so(4)) = \left(\mathbf{L}^2 + \mathbf{A}^2\right)$$

$$C_2'(so(4)) = \mathbf{L} \cdot \mathbf{A} \quad. \tag{15.26}$$

After lengthy manipulations, the Hamiltonian operator can be written as

$$H = -\frac{1}{2\left(C_2(so(4)) + 1\right)}. \tag{15.27}$$

As one can see from the definition (15.23) of the Runge-Lenz vector, the scalar product of $\mathbf{L}$ and $\mathbf{A}$ vanishes, $\mathbf{L} \cdot \mathbf{A} = 0$. The second invariant operator does not appear therefore in H. The basis states of $so(4)$ are labelled by two quantum numbers $[\omega_1, \omega_2]$. However, due to the vanishing of $C_2'(so(4))$, only the totally symmetric representations $[\omega] \equiv [\omega, 0]$ appear. The energy eigenvalues can be obtained from the eigenvalues of the Casimir operators in Chap. 7. They are given by

$$E(\omega) = -\frac{1}{2(\omega(\omega + 2) + 1)} \qquad \omega = 0, 1, \ldots, \infty. \tag{15.28}$$

In order to label completely the states one needs to consider the branching $so(4) \supset so(3) \supset so(2)$. This branching was considered in Chap. 9. The complete labelling of states is $|\omega, l, m\rangle$, with branching $l = \omega, \omega - 1, \ldots, 0$ and $m = -l, \ldots, +l$. It is customary to introduce the "principal quantum number" $n = \omega + 1$. In terms of this quantum number the energy levels are

$$E(n, l, m) = -\frac{1}{2n^2}. \tag{15.29}$$

This is the celebrated Bohr formula that gives the energy levels of the non-relativistic hydrogen atom. The corresponding energy level diagram is shown in Fig. 15.3. The states $|n, l, m\rangle$ are degenerate with total degeneracy n^2. Therefore $so(4)$ is the degeneracy algebra of the Coulomb problem in $\nu = 3$ dimensions. A detailed account is given in Wybourne (1974).

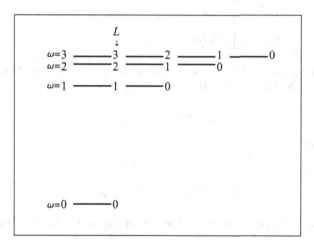

Fig. 15.3 The spectrum of the three-dimensional Coulomb problem. States up to $\omega = 3$ are shown. The principal quantum number is $n = \omega + 1$

Example 4. Coulomb problem in six dimensions

This problem and its application to the three-body problem is best solved in hyperspherical Jacobi coordinates. One first introduces two vectors $\boldsymbol{\rho}$ and $\boldsymbol{\lambda}$ or equivalently $\rho, \vartheta_\rho, \varphi_\rho$ and $\lambda, \vartheta_\lambda, \varphi_\lambda$. The Jacobi hyperspherical coordinates are $r, \xi, \vartheta_\rho, \varphi_\rho, \vartheta_\lambda, \varphi_\lambda$ with

$$r = \left(\rho^2 + \lambda^2\right)^{1/2} \quad \xi = \arctan\left(\rho/\lambda\right). \tag{15.30}$$

Introducing the notation

$$q_i \equiv (\boldsymbol{\rho}, \boldsymbol{\lambda}) \quad p_i \equiv (\mathbf{p}_\rho, \mathbf{p}_\lambda) \quad i = 1, \dots, 6, \tag{15.31}$$

one can construct the 15 elements of the Lie algebra $so(6)$ by

$$L_{ij} = q_i p_j - q_j p_i, \quad i, j = 1, \dots, 6 \quad i < j. \tag{15.32}$$

To these, one can add the six components of the normalized Runge-Lenz vector

$$A_i = \frac{1}{\sqrt{-2H}} \left[\frac{1}{2} \left(L_{ij} p_j - p_j L_{ji} \right) - \frac{q_i}{r} \right] \quad i = 1, \dots, 6. \tag{15.33}$$

The 21 elements L_{ij}, A_i satisfy the commutation relations

$$\begin{aligned}
\left[L_{ij}, L_{kl} \right] &= i \left(\delta_{ik} L_{jl} + \delta_{jl} L_{ik} - \delta_{il} L_{jk} - \delta_{jk} L_{il} \right) \\
\left[A_i, A_j \right] &= i L_{ij} \\
\left[L_{ij}, A_k \right] &= i \left(\delta_{ik} A_j - \delta_{jk} A_i \right).
\end{aligned} \tag{15.34}$$

These commutation relations are isomorphic to those of $so(7)$. The 21 elements also commute with the Hamiltonian H

$$\left[L_{ij}, H \right] = [A_i, H] = 0 \tag{15.35}$$

where

$$H = \frac{1}{2} \left(\mathbf{p}_\rho^2 + \mathbf{p}_\lambda^2 \right) - \frac{1}{r}. \tag{15.36}$$

The algebra $so(7)$ possesses three Casimir operators C_2, C_4, C_6. The quadratic operator is

$$C_2(so(7)) = \sum_{i<j}^6 L_{ij}^2 + \sum_i^6 A_i^2. \tag{15.37}$$

After some lengthy manipulations the Hamiltonian can be written as

$$H = -\frac{1}{2\left[C_2(so(7)) + \frac{25}{4}\right]}. \tag{15.38}$$

The expectation value of this Hamiltonian in the representation $[\omega] \equiv [\omega, 0, 0]$ is

$$E(\omega) = -\frac{1}{2\left[\omega(\omega + 5) + \frac{25}{4}\right]}, \qquad \omega = 0, 1, \ldots, \infty. \tag{15.39}$$

The representations of $so(7)$ that appear are the totally symmetric representations $[\omega]$, in view of conditions analogous to $\mathbf{L} \cdot \mathbf{A} = 0$ in the three dimensional case. In order to label completely the states, one needs to study the branching of representations of $so(7)$. This can be done using the techniques of Chap. 6. The labelling of states appropriate for the choice of coordinates in this example is $so(7) \supset so(6) \supset so_\rho(3) \oplus so_\lambda(3) \supset so(3) \supset so(2)$ with quantum numbers

$$\left| \begin{array}{ccccc} so(7) \supset & so(6) \supset & so(3) \oplus so(3) \supset & so(3) \supset & so(2) \\ \downarrow & \downarrow & \downarrow & \downarrow & \downarrow \\ \omega & \gamma & l_\rho, l_\lambda & L & M \end{array} \right). \tag{15.40}$$

Here $\gamma = \omega, \omega - 1, \ldots, 1, 0$, and the values of l_ρ, l_λ are obtained by partitioning γ as $\gamma = 2n_\gamma + l_\rho + l_\lambda, n_\gamma = 0, 1, \ldots$. The values of L are obtained from $|l_\rho + l_\lambda| \geq L \geq |l_\rho - l_\lambda|$ and $M = -L, \ldots, +L$ as usual. The energy level diagram of the six dimensional Coulomb problem is shown in Fig. 15.4. The spectrum is degenerate and $so(7)$ is the degeneracy algebra. The Coulomb problem in six dimensions is of interest in the three-body problem, in particular the three quark system in hadronic physics (Santopinto et al. 1995).

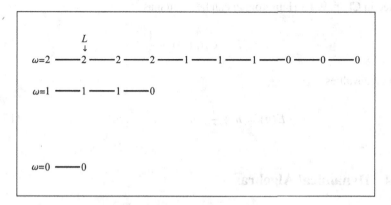

Fig. 15.4 The spectrum of the six-dimensional Coulomb problem. States up to $\omega = 2$ are shown

Table 15.1 Exactly solvable quantum mechanical problems in one dimension

Case	$V(x)$	$E(n)$
(I)	$\frac{1}{2}\frac{\kappa(\kappa-1)}{\sin^2 x}$	$\frac{1}{2}(\kappa + 2n)^2$
(II)	$-\frac{1}{2}\frac{\kappa(\kappa-1)}{\cosh^2 x}$	$-\frac{1}{2}(\kappa - 1 - n)^2$
(III)	$\kappa^2\left(e^{-2x} - 2e^{-x}\right)$	$-\kappa^2 + \frac{1}{2}\left[2\sqrt{2}\kappa(n + \frac{1}{2}) - (n + \frac{1}{2})^2\right]$
(IV)	$\frac{1}{2}\kappa^2 x^2$	$\kappa\left(n + \frac{1}{2}\right)$

15.3 Degeneracy Algebra in $\nu = 1$ Dimension

The case $\nu = 1$ is a special case, since the degeneracy algebra of the one-dimensional Hamiltonian, in units $\hbar = m = 1$,

$$H = -\frac{1}{2}\frac{d^2}{dx^2} + V(x) \tag{15.41}$$

is always the trivial algebra $u(1) \sim so(2)$. Because in one dimension there is no rotational invariance to impose, the class of exactly solvable problems in $\nu = 1$ dimension, that is of problems that can be written in terms of Casimir operators of an algebra g, is much wider than that in $\nu \geq 2$ dimensions. A partial list is given in Table 15.1. The Hamiltonian for these problems is either linear, case IV, or quadratic, cases I–III, in the Casimir operators and hence the eigenvalues, $E(n)$, are either linear or quadratic in the quantum number n. The strength of the potential is given in the table either by $\kappa(\kappa - 1)$ or by κ^2.

Example 5. The one dimensional harmonic oscillator

The one dimensional harmonic oscillator, $V(x) = \frac{1}{2}kx^2$, in units $k = 1$, has been treated in Chap. 9. Its Hamiltonian can be written as

$$H = C_1(u(1)) + \frac{1}{2} \tag{15.42}$$

with eigenvalues

$$E(n) = n + \frac{1}{2} \quad n = 0, 1, \ldots, \infty. \tag{15.43}$$

15.4 Dynamical Algebras

The degeneracy algebra g_d allows one to solve the eigenvalue problem and thus classify the degenerate multiplets. However, in the cases discussed in the previous

Chap. 14 all states of the system were assigned to an irreducible representation of an algebra g, and all operators $H, T, ..$ were written in terms of elements of g. It is of interest to do the same for quantum mechanical problems. This problem is often called the *embedding* problem. The algebra g that contains the degeneracy algebra $g \supset g_d$ is called the dynamical algebra. Included in the dynamical algebra g there are now raising and lowering operators that relate the different representations of g. The embedding problem does not have a unique solution. An additional complication is that often the quantum mechanical problem has an infinite number of bound ($E < 0$) eigenstates, as in the two cases of Sect. 15.2. The dynamical algebra must therefore be either a non-compact algebra or obtained by a limiting process (the contraction process discussed in Chap. 1, Sect. 1.17) from a compact algebra. Both of these problems are outside the scope of this book and will therefore be only mentioned here.

15.5 Dynamical Algebras in $\nu \geq 2$ Dimensions

15.5.1 Harmonic Oscillator

The degeneracy algebra is $u(\nu)$. A commonly used dynamical algebra is the symplectic algebra $sp(2\nu, R) \supset u(\nu)$. (Hwa and Nuyts, 1966) However, this embedding has the disadvantage that two irreducible representations are needed to accommodate all the states of the harmonic oscillator. A simpler embedding is obtained by drawing on the results of Chap. 9. Introducing a fictitious coordinate, s, and momentum, $p_s = \frac{1}{i}\frac{d}{ds}$, one can construct the algebra $u(\nu + 1)$. The algebra $u(\nu + 1) \supset u(\nu)$ can be used as a dynamical algebra of the harmonic oscillator. All states are assigned to the representation $[N]$ of $u(\nu + 1)$ with $N \to \infty$. The degenerate multiplets are labelled by $n = 0, 1, \ldots, N \to \infty$.

15.5.2 Coulomb Problem

The degeneracy algebra is $so(\nu + 1)$. The non-compact algebras $so(\nu + 1, 2) \supset so(\nu + 1, 1) \supset so(\nu + 1)$ have been used as dynamical algebras (Bacry 1966; Sudarshan et al. 1965; Barut 1969). However, the representation theory of non-compact algebras is rather complicated and requires special attention. Again, a simpler embedding is obtained by introducing a fictitious coordinate s and momentum $\frac{1}{i}\frac{d}{ds}$, and constructing the algebra $so(\nu + 2)$. This algebra, instead of $so(\nu + 1, 1)$, can be used as dynamical algebra of the Coulomb problem in $\nu \geq 2$ dimensions. All states are assigned to a representation $[\Gamma]$ of $so(\nu + 2)$ with $\Gamma \to \infty$. The degenerate multiplets are labelled by $\omega = 0, 1, \ldots$, $\Gamma \to \infty$.

15.6   Dynamical Algebra in $v = 1$ Dimension

The algebra $u(2)$ has been used extensively as a dynamical algebra in $v = 1$ dimension. In particular, this algebra has been used in the study of two quantum mechanical problems of practical interest, already listed in Sect. 15.3, $V(x) = -\frac{k}{\cosh^2 x}$ (called the Pöschl-Teller potential) and $V(x) = k(e^{-2x} - 2e^{-x})$ (called the Morse potential).

15.6.1   Pöschl-Teller Potential

The Hamiltonian operator for this problem in units $\hbar = m = 1$ is

$$H = -\frac{1}{2}\frac{d^2}{dx^2} - \frac{k}{\cosh^2 x}, \tag{15.44}$$

where x is a dimensionless coordinate. In order to construct the dynamical algebra $u(2) \supset su(2)$, introduce two variables ϑ and φ and consider the differential realization of $su(2) \sim so(3)$ on the sphere, given in Chap. 11,

$$I_z = -i\frac{\partial}{\partial\varphi}$$

$$I_\pm = -ie^{\mp i\varphi}\left(\frac{\partial}{\partial\vartheta} \mp i\cot\vartheta\frac{\partial}{\partial\varphi}\right) \tag{15.45}$$

and

$$I^2 = -\left[\frac{1}{\sin\vartheta}\frac{\partial}{\partial\vartheta}\left(\sin\vartheta\frac{\partial}{\partial\vartheta}\right) + \frac{1}{\sin^2\vartheta}\frac{\partial^2}{\partial\varphi^2}\right]. \tag{15.46}$$

The simultaneous eigenfunctions of I^2 and I_z

$$I^2\chi_j^m = j(j+1)\chi_j^m$$
$$I_z\chi_j^m = m\chi_j^m \tag{15.47}$$

with j and m integer, are

$$\chi_j^m(\vartheta,\varphi) = u_j^m(\vartheta)\,e^{im\varphi} \tag{15.48}$$

where $u_j^m(\vartheta)$ satisfies the equation

$$\left[-\frac{1}{\sin\vartheta}\frac{\partial}{\partial\vartheta}\left(\sin\vartheta\frac{\partial}{\partial\vartheta}\right) + \frac{m^2}{\sin^2\vartheta} \right] u_j^m(\vartheta) = j(j+1)u_j^m(\vartheta). \tag{15.49}$$

The solutions of (15.49) are the associated Legendre functions $P_j^m(\cos\vartheta)$ and the functions $\chi_j^m(\vartheta,\varphi)$ are just the spherical harmonics $Y_{jm}(\vartheta,\varphi)$. The substitution

$$\cos\vartheta = \tanh x \qquad -\infty < x < +\infty \tag{15.50}$$

brings (15.49) to the form

$$\left[-\frac{d^2}{dx^2} - \frac{j(j+1)}{\cosh^2 x} \right] u_j^m(x) = -m^2 u_j^m(x). \tag{15.51}$$

This is, apart from a factor 2, the Schrödinger equation with Pöschl-Teller potential. The strength of the potential k is related to the eigenvalue of the Casimir operator of $su(2) \sim so(3)$, $j(j+1)$, by $k = \frac{1}{2}j(j+1)$. The Hamiltonian operator can be written as

$$H = -\frac{1}{2}I_z^2 \tag{15.52}$$

with eigenvalues

$$E(m) = -\frac{1}{2}m^2. \tag{15.53}$$

In the bra-ket notation, the eigenfunctions can be written as

$$\left| \begin{matrix} u(2) \supset so(2) \\ \downarrow \qquad \downarrow \\ j \qquad m \end{matrix} \right\rangle. \tag{15.54}$$

All eigenstates are assigned to the representation j of $u(2)$. This algebra is the dynamical algebra of the problem. The Hamiltonian is written in terms of elements of this algebra, in fact in terms of the Casimir operator I_z of the subalgebra $so(2)$, which is the degeneracy algebra of the problem. There is a peculiarity in this case due to the fact that the algebra $so(2)$ is an orthogonal algebra in an even number of dimensions and thus $m = -j, \ldots, +j$. Since the eigenvalues depend only on m^2, they are double degenerate. They correspond to the eigenvalues of the potential and its reflection. The algebra $u(2)$ should therefore be used as dynamical algebra with the proviso that only states with $m \geq 0$ should be considered. The spectrum of the Pöschl-Teller potential with this proviso is shown in Fig. 15.5. The Pöschl-Teller potential is of interest in molecular physics (Iachello and Oss 1993). In applications in molecular physics, often the quantum number $m = j, j-1, \ldots, 1, 0$ is replaced by the so-called vibrational quantum number $v = j - m, v = 0, 1, \ldots, j$. In contrast

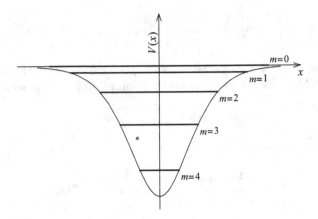

Fig. 15.5 The spectrum of the one-dimensional Pöschl-Teller potential superimposed to the potential. The strength of the potential is characterized by $j = 4$

with the harmonic oscillator of Example 6, the Pöschl-Teller potential has a finite number of bound states. Thus $j = $ finite, and the representations of $u(2)$ that appear in this problem are the usual finite dimensional representations.

15.6.2 Morse Potential

The Hamiltonian for this problem in units $\hbar = m = 1$ is

$$H = -\frac{1}{2}\frac{d^2}{dx^2} + k\left[e^{-2x} - 2e^{-x}\right]. \tag{15.55}$$

In order to construct the spectrum generating algebra, introduce again two variables s and t. A realization of $u(2)$ in terms of differential operators $s\frac{\partial}{\partial s}, s\frac{\partial}{\partial t}, t\frac{\partial}{\partial s}, t\frac{\partial}{\partial t}$ is

$$\hat{F}_x = \frac{1}{2}\left(st - \frac{\partial^2}{\partial s\partial t}\right)$$

$$\hat{F}_y = \frac{1}{2i}\left(s\frac{\partial}{\partial t} - t\frac{\partial}{\partial s}\right)$$

$$\hat{F}_z = \frac{1}{4}\left(s^2 - t^2 - \frac{\partial^2}{\partial s^2} + \frac{\partial^2}{\partial t^2}\right)$$

$$\hat{N} = \frac{1}{2}(s^2 + t^2 - \frac{\partial^2}{\partial s^2} - \frac{\partial^2}{\partial t^2} - 2). \tag{15.56}$$

This realization is the oscillator realization of Chap. 9, Sect. 9.3 written in terms of differential operators. A hat is put in (15.56) to distinguish an operator from its eigenvalue. Consider now the simultaneous eigenstates of $\hat{N}$ and $\hat{F}_y$. Introducing polar coordinates

$$s = r\cos\varphi \quad 0 \le r < \infty$$
$$t = r\sin\varphi \quad 0 \le \varphi < 2\pi, \tag{15.57}$$

the two operators can be rewritten as

$$\hat{F}_y = -\frac{i}{2}\frac{\partial}{\partial\varphi}$$

$$\hat{N} = \frac{1}{2}\left(r^2 - \frac{1}{r}\frac{\partial}{\partial r}r\frac{\partial}{\partial r} - \frac{1}{r^2}\frac{\partial^2}{\partial\varphi^2}\right) - 1. \tag{15.58}$$

The simultaneous eigenstates of $\hat{N}$ and $\hat{F}_y$ can be written as

$$\psi_{N,m_y}(r,\varphi) = R_{N,m_y}(r)e^{2im_y\varphi} \tag{15.59}$$

where $2m_y = $ integer since ψ should be periodic in φ with period 2π, and $R_{N,m_y}(r)$ satisfies

$$\frac{1}{2}\left(-\frac{1}{r}\frac{\partial}{\partial r}r\frac{\partial}{\partial r} + \frac{4m_y^2}{r^2} + r^2\right)R_{N,m_y}(r) = (N+1)R_{N,m_y}(r). \tag{15.60}$$

By the change of variable

$$r = \sqrt{N+1}\,e^{-x/2} \tag{15.61}$$

this equation can be brought to the form

$$\left[-\frac{d^2}{dx^2} + \left(\frac{N+1}{2}\right)^2(e^{-2x} - 2e^{-x})\right]R_{N,m_y}(x) = -m_y^2 R_{N,m_y}(x). \tag{15.62}$$

This is, apart from a factor 2, the Schrödinger equation with Morse potential. The strength of the potential k is related to the eigenvalue of the Casimir operator of $su(2)$ by

$$k = \frac{1}{2}\left(\frac{N+1}{2}\right)^2. \tag{15.63}$$

The Hamiltonian operator can be written as

$$H = -\frac{1}{2}\hat{F}_y^2 \tag{15.64}$$

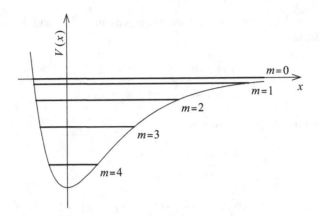

Fig. 15.6 The spectrum of the one-dimensional Morse potential. The strength of the potential is characterized by $F = \frac{N}{2} = 4$

with eigenvalues

$$E(m_y) = -\frac{1}{2}m_y^2. \tag{15.65}$$

In bra-ket notation, the eigenfunctions can be written as

$$\left| \begin{array}{cc} u(2) \supset u(1) \\ \downarrow \quad \downarrow \\ F \quad m_y \end{array} \right\rangle, \tag{15.66}$$

where, as discussed in Chap. 9, Sect. 9.3, $F = \frac{N}{2}$. The values of m_y are $m_y = -\frac{N}{2}, \ldots, +\frac{N}{2} = -F, \ldots, +F$, but, for reasons given above, only $m_y \geq 0$ need be considered. The spectrum of the Morse potential is shown in Fig. 15.6. The Morse potential is of great interest in molecular physics (Alhassid et al. 1983). Here also it is customary to introduce a vibrational quantum number $v = \frac{N}{2} - m_y, v = 0, 1, \ldots, \frac{N}{2}$ or $\frac{N-1}{2}$ (N =even or odd), and the number of bound states is finite. The Morse potential problem in one dimension is solvable not only for the values such that N integer, but for any strength k. The general solution requires however the use of *projective* representations of $u(2)$ rather than tensor representations. Projective representations of $u(2)$ (Bargmann, 1947) are outside the scope of these lectures note and will not be discussed.

The Morse potential has also been associated with representations of the non-compact algebra $su(1, 1)$ (Cordero and Hojman, 1970).

15.6.3 Lattice of Algebras

The concept of lattice of algebras introduced in Chap. 9, can be used here as well. One dimensional exactly solvable problems are characterized by the lattice of algebras

$$
\begin{array}{c}
u(2) \\
| \\
u(1) \sim so(2)
\end{array}
\tag{15.67}
$$

Here $u(2)$ is the dynamical algebra and $u(1) \sim so(2)$ is the degeneracy algebra. Because of the isomorphism $u(1) \sim so(2)$, all exactly solvable problems in $\nu = 1$ dimension have the same structure. In particular, the two problems discussed in Sects. 6.1 and 6.2 have the same bound state spectrum

$$
E(m) = -\frac{1}{2}m^2.
\tag{15.68}
$$

Problems with the same bound ($E < 0$) spectrum are called *isospectral*. (They differ, however, in the scattering ($E > 0$) spectrum, not discussed here.)

The algebra $u(2)$ can also be used as a dynamical algebra of the harmonic oscillator of Sect. 15.3. The Hamiltonian is now linear in the Casimir operator C_1 of $u(1)$ with eigenvalue

$$
E(n) = n + \frac{1}{2}.
\tag{15.69}
$$

The harmonic oscillator has an infinite number of bound states and thus $N \to \infty$ and $n = 0, 1, \ldots, \infty$.

References

Alhassid, Y., Iachello, F., Gürsey, F.: Group theory of the Morse oscillator. Chem. Phys. Lett. **99**, 27 (1983)

Bacry, H.: The de Sitter group $L_{4,1}$ and the bound states of the hydrogen atom. Nuovo Cimento **41A**, 222 (1966)

Baird, G.E., Biedenharn, L.C.: On the representations of semisinple Lie groups-II. J. Math. Phys. **4**, 1449 (1963)

Bargmann, V.: Zur Theorie des Wassenstoffatoms. Z. Phys. **99**, 576 (1936)

Bargmann, V.: Irreducible unitary representations of the Lorentz group. Ann. Math. **48**, 568 (1947)

Barut A.O.: Dynamical groups and their currents. Springer Tracts Mod. Phys. **50**, 1 (1969)

Barut, A.O., Böhm, A.: Dynamical groups and mass formulas. Phys. Rev. B **139**, 1107 (1965)

Barut, A.O., Rączka, R.: Theory of Group Representations and Applications. World Scientific, Singapore (1986)

Berezin, F.A.: An Introduction to Superanalysis. D. Riedel, Dordrecht (1987)

Bijker, R., Iachello, F., Leviatan, A.: Algebraic models of hadronic structure. I. Non strange baryons. Ann. Phys. (NY) **236**, 69 (1994a)

Bijker, R., Leviatan, A.: Algebraic treatment of collective excitations in baryon spectroscopy. In: B. Gruber, B., Otsuka, T., (eds.) Symmetry in Science VII: Spectrum Generating Algebras and Dynamic Symmetries in Physics. Plenum Press, New York (1994b)

Bröcker, T., tom Dieck, T.: Representations of Compact Lie Groups. Springer, New York (1985)

Caprio, M.A.: Applications of the coherent state formalism to multiply excited states. J. Phys. A **38**, 6385 (2005)

Cartan, E.: The Theory of Spinors. Hermann, Paris (1966)

Cartan, E.: Sur la Structure des Groupes de Transformation Finis and Continus. Thése, Paris (1894)

Cartan, E.: Sur une classe remarquable d'espaces de Riemann. Bull. Soc. Math. France **54**, 214 (1926); **55**, 114 (1927)

Casimir, H.: Über die Konstruktion einer zu den irreduzibelen Darstellung halbeinfacher kontinuerlichen Gruppen gehörigen Differential-gleichung. Proc. R. Akad. Amst. **34**, 844 (1931)

Cejnar, P., Iachello, F.: Phase structure of interacting boson models in arbitrary dimensions. J. Phys. A **40**, 581 (2007)

Chaichian, M., Hagedorn, R.: Symmetry in Quantum Mechanics: From Angular Momentum to Supersymmetry. IOP, Philadelphia (1998)

Chen, J.-Q.: Group Representation Theory for Physicists. World Scientific, Singapore (1989)

Chen, J.-Q., Chen, B.-Q., Klein, A.: Factorization for commutators: the Wick theorem for coupled operators. Nucl. Phys. A **554**, 61 (1993)

© Springer-Verlag Berlin Heidelberg 2015

F. Iachello, *Lie Algebras and Applications*, Lecture Notes in Physics 891,
DOI 10.1007/978-3-662-44494-8

Cordero, P., Hojman, S.: Algebraic solution of a short-range potential problem. Lett. Nuovo Cimento **4**, 1123 (1970)

Cornwell, J.F.: Group Theory in Physics: An Introduction. Academic Press, San Diego (1997)

De Shalit, A., Talmi, I.: Nuclear Shell Theory. Academic Press, New York (1963)

de Swart, J.J.: The Octet Model and its Clebsch-Gordan Coefficients. Rev. Mod. Phys. **35**, 916 (1963)

Dothan, Y., Gell-Mann, M., Ne'eman, Y.: Series of hadron levels as representations of non-compact groups. Phys. Lett. **17**, 148 (1965)

Duistermaat, J.J., Kolk, J.A.C.: Lie Groups. Springer, Berlin (2000)

Dynkin, E.B.: The structure of semisimple Lie algebras. Usp. Mat. Nauk (N.S.) **2**, 59 (1947). Translated in Am. Math. Soc. Transl. (1) **9**, 308 (1962)

Elliott, J.P.: Collective motion in the nuclear shell model. I. Classification scheme for states of mixed configurations. Proc. R. Soc. **A245**, 128 (1958)

Flowers, R.H.: Studies in j–j coupling. I. Classification of nuclear and atomic states. Proc. R. Soc. **A212**, 248 (1952)

Fock, V.A.: Zur theorie des Wassenstoffatoms. Z. Phys. **98**, 145 (1935)

Frank, A., van Isacker, P.: Algebraic Methods in Molecular and Nuclear Structure Physics. Wiley, New York (1994)

Franzini, P., Radicati, L.A.: On the validity of the supermultiplet model. Phys. Lett. **6**, 32 (1963)

French, J.B.: Multipole and sum-rule methods in spectroscopy. In: Bloch, C., (ed.) Proceedings of the International School of Physics "Enrico Fermi", Course XXXVI, p. 278. Academic Press, New York (1966)

Fulton, W., Harris, J.: Representation Theory: A First Course. Springer, New York (1991)

Gel'fand, I.M., Cetlin, M.L.: Finite-dimensional representations of a group of unimodular matrices. Dokl. Akad. Nauk SSSR **71**, 8 & 825 (1950)

Gel'fand, I.M., Cetlin, M.L.: Finite-dimensional representations of groups of orthogonal matrices. Dokl. Akad. Nauk SSSR **71**, 1017 (1950)

Gell-Mann, M.: Symmetries of baryons and mesons. Phys. Rev. **125**, 1067 (1962)

Georgi, H.: Lie Algebras in Particle Physics. Perseus Books, Reading (1962)

Gilmore, R.: Lie Groups, Lie Algebras and Some of Their Applications. Wiley, New York (1974)

Goldstein, H., Poole, C., Safko, J.: Classical Mechanics. Addison-Wesley, San Francisco (2002)

Gürsey, F., Radicati, L.A.: Spin and unitary spin independence of strong interactions. Phys. Rev. Lett. **13**, 173 (1964)

Hall, B.C.: Lie Groups, Lie Algebras, and Representations. Springer, New York (2003)

Hamermesh, M.: Group Theory and its Applications to Physical Problems. Addison-Wesley, Reading (1962)

Hermann, R.: Lie Groups for Physicists. W.A. Benjamin, New York (1966)

Hladik, J.: Spinors in Physics. Translated from the French edition Spineurs en Physique. Springer, New York (1999)

Humphreys, J.E.: Introduction to Lie Algebras and Representation Theory, Graduate Text in Mathematics. Springer, New York (1972)

Hwa, R.C., Nuyts, T.: Group embedding for the harmonic oscillator. Phys. Rev. **145**, 1188 (1966)

Iachello, F.: Dynamic symmetries in nuclei. In: Böhm, A. (ed.) Group Theoretical Methods in Physics. Lange Springer, Berlin (1979)

Iachello, F.: Algebraic theory of the three-body problem. In: Gruber, B., (ed.) Symmetries in Science VIII. Plenum Press, New York (1995)

Iachello, F., Arima, A.: The Interacting Boson Model. Cambridge University Press, Cambridge (1987)

Iachello, F., Levine, R.D.: Algebraic Theory of Molecules. Oxford University Press, Oxford (1995)

Iachello, F., van Isacker, P.: The Interacting Boson Fermion Model. Cambridge University Press, Cambridge (1991)

Iachello, F., Oss, S.: Algebraic Model of Bending Vibrations of Complex Molecules. Chem. Phys. Lett. **205**, 285 (1993)

Iachello, F., Oss, S.: Algebraic approach to molecular spectra: two-dimensional problems. J. Chem. Phys. **104**, 6956 (1996)

Jacobson, N.: Lie Algebras. Dover, New York (1979). Reprinted from General Publishing Company, Toronto, Ontario, Canada (1962)

Jauch, J.M., Hill, E.H.: On the problem of degeneracy in quantum mechanics. Phys. Rev. **57**, 641 (1940)

Kirillov, A.A.: Elementy Teorii Predstavlenii. Nauka, Moscow (1972). Translated in Elements of the Theory of Representation. Springer, Berlin (1976)

Kirillov, A., Jr.: An Introduction to Lie Groups and Lie Algebras. Cambridge University Press, Cambridge (2008)

Killing, W.: Die Zusammensetzung der Stetigen Endlichen Transformationgruppen, I–IV. Math. Ann. **31**, 252 (1888); **33**, 1 (1889a); **34**, 57 (1889b); **36**, 161 (1890)

Lie, S., Scheffers, G.: Vorlesungen über Kontinuerliche Gruppen. Leipzig (1893)

Lipkin, H.J.: Lie Groups for Pedestrians. Dover, Mineola (2002). Reprinted from North-Holland Publishing Company, Amsterdam (1966)

Messiah, A.: Quantum Mechanics. Wiley, New York. Translated from the French edition, Dunod, Paris (1958)

Okubo, S.: Note on unitary symmetry in strong interactions. Prog. Theor. Phys. **27**, 949 (1962)

Pauli, W.: Über das Wassenstoffspektrum von Standpunkt der neuen Quantenmechanik. Z. Phys. **36**, 336 (1926)

Perelomov, A.M.: Generalized Coherent States and Their Applications. Springer, Berlin (1986)

Perelomov, A.M., Popov, V.S.: Casimir Operators for u(n) and su(n). Sov. J. Nucl. Phys. **3**, 676 (1966a)

Perelomov, A.M., Popov, V.S.: Casimir operators of orthogonal and symplectic groups. Sov. J. Nucl. Phys. **3**, 819 (1966b)

Racah, G.: Group Theory and Spectroscopy, Mimeographed Lecture Notes. Princeton, New Jersey (1951). Reprinted in Springer Tracts in Modern Physics **37**, 28 (1965)

Racah, G.: Theory of complex spectra-IV. Phys. Rev. **76**, 1352 (1949)

Santopinto, E., Giannini, M., Iachello, F.: Algebraic spproach to the hyper-coulomb problem. In: Gruber, B., (ed.) Symmetry in Science VIII. Plenum Press, New York (1995)

Sattinger, D.H., Weaver, O.L.: Lie Groups, Lie Algebras with Applications to Physics, Geometry, and Mechanics. Springer, New York (1986)

Schwinger, J.: On angular momentum. In: Biedenharn, L.C., van Dam, H., (eds.) Quantum Theory of Angular Momentum, p. 229. Academic Press, New York (1965)

Serre, J-P.: Lie Algebras and Lie Groups. W.A. Benjamin, New York (1965)

Slansky, R.: Group theory for unified model building. Phys. Rep. **79**, 1 (1981)

Sudarshan, E.C.G., Mukunda, N., O'Raifeartaigh, L.: Group theory of the Kepler problem. Phys. Lett. **19**, 322 (1965)

van der Waerden, B.L.: Die Gruppentheoretische Methode in der Quantenmechanik. Math. Zeitschr. **37**, 446 (1933)

van Roosmalen, O.S.: Algebraic Description of Nuclear and Molecular Rotation-Vibration Spectra. Ph.D. Thesis, University of Groningen, The Netherlands (1982)

Varadarajan, V.S.: Lie Groups, Lie Algebras, and Their Representations. Springer, New York (1984)

Weyl, H.: Theorie der Darstellung kontinuerlichen halbeinfacher Gruppen durch lineare Transformationen, I–III. Math. Zeitschr. **23**, 271 (1925); **24**, 328 & 377 (1926)

Wybourne, B.G.: Classical Groups for Physicists. Wiley, New York (1974)

Zhang, W., Feng, D.H., Gilmore, R.: Coherent states: theory and some applications. Rev. Mod. Phys. **62**, 867 (1990)

Index

© Springer-Verlag Berlin Heidelberg 2015

F. Iachello, *Lie Algebras and Applications*, Lecture Notes in Physics 891,
DOI 10.1007/978-3-662-44494-8